Understanding Experimental Planning for Advanced Level Chemistry

The Learner's Approach

Understanding Experimental Planning for Advanced Level Chemistry

The Learner's Approach

Kim Seng Chan
BSc (Hons), PhD, PDGE (Sec), MEd, MA (Ed Mgt), MEd (G Ed), MEd (Dev Psy)
Victoria Junior College, Singapore

Jeanne Tan
BSc (Hons), PDGE (Sec), MEd (LST)

Published by

World Scientific Publishing Co. Pte. Ltd.
5 Toh Tuck Link, Singapore 596224
USA office: 27 Warren Street, Suite 401-402, Hackensack, NJ 07601
UK office: 57 Shelton Street, Covent Garden, London WC2H 9HE

Library of Congress Cataloging-in-Publication Data
Chan, Kim Seng.
  Understanding experimental planning for advanced level chemistry : the learner's approach / Kim Seng Chan, Jeanne Tan.
    pages cm
  Includes index.
  ISBN 978-9814667906 (pbk. : alk. paper) -- ISBN 9814667900 (pbk. : alk. paper)
  1. Chemistry--Laboratory manuals. I. Tan, Jeanne. II.Title.
  QD45.C35 2015
  540--dc23
                       2015008834

British Library Cataloguing-in-Publication Data
A catalogue record for this book is available from the British Library.

Copyright © 2015 by World Scientific Publishing Co. Pte. Ltd.

All rights reserved. This book, or parts thereof, may not be reproduced in any form or by any means, electronic or mechanical, including photocopying, recording or any information storage and retrieval system now known or to be invented, without written permission from the publisher.

For photocopying of material in this volume, please pay a copying fee through the Copyright Clearance Center, Inc., 222 Rosewood Drive, Danvers, MA 01923, USA. In this case permission to photocopy is not required from the publisher.

Printed in Singapore by Mainland Press Pte Ltd.

PREFACE

In a normal chemical laboratory, students need to perform chemical experiments to solve some problems. It is through these experiments that we hope to introduce to our students, the basics of physical observation and to the scientific method. We also hope to familiarize our students with several basic quantitative methods such as volumetric and gravimetric measurements, care in handling glassware during the experimentation and safety awareness to the use and handling of chemicals. Lastly, based on the observational information that they have acquired through the experimentation, we hope for them to verify empirically the chemical theory they have acquired in the classroom both in a qualitative and quantitative manner.

For each experiment that a student performs, it normally comprises an introduction, an experimental procedure, tables for data tabulation and results, some questions that link up with the theoretical concepts and space for sample calculations. Usually, it is not a major issue for students to diligently follow the procedure to obtain the necessary data. However, from our years of experience, many students do not understand the rationale behind the procedural steps, the quantities and the apparatus used. Thus, when students are asked to plan an experiment to solve a problem, e.g., using a titration or gravimetric method, they are usually at a loss as to how to start off with the planning. It is therefore the intention of this book to share with our students how experimental planning can be done without much hassle.

1.1 How to Start Off with the Planning?

A context would usually be set for the problem; importantly, one must determine the aim or objective for the experiment from the given context. For example:

> "A student obtained some impure limestone, which is an impure form of calcium carbonate, $CaCO_3$, during a field trip. Describe how you would help the student to determine the percentage of $CaCO_3$ in a sample of the impure limestone by a method using titration. Take note that limestone does not dissolve in water."

Thus, based on the above scenario, the aim of the experiment is to "determine the percentage of $CaCO_3$ in a sample of the impure limestone". The method that should be used is the titration method. But what is the chemical that would react with the carbonate? From this, the student needs to know that an acid can react with the carbonate. Since we are going to use a titration method, then, the next question is: can we simply add acid from the burette to the carbonate? Of course we cannot; as the limestone does not dissolve in water, you would not know when to stop adding the acid. And when would the end-point be? These questions should form a coherent thought process of the student during the planning stage. But once he knows that he can dissolve the limestone in an excess known amount of acid, he can simply titrate the unused acid. But other issues such as: how much of the sample to use? What is the concentration of the acid? What is the indicator used? How much to pipette? Should dilution be done before titration? All these questions need to be asked and answered during the planning stage. We would discuss them in further detail for each of the specific types of experiment that are typically questioned during the GCE 'A' level examination.

1.2 What Is Considered a Good Plan?

A well-planned experiment would give you the best result. A properly planned experiment takes into consideration health and safety issues. By going through the planning stage, the students who know their roles can help to prevent any hazard arising due to lack of understanding. What do we mean? It is through the planning stage that students become more aware of the safety issues of the chemicals and the apparatus that they are using. This would enable the students to feel more assured of the safety precautions, and fewer accidents are likely to happen.

A properly planned experiment clearly maps out the steps that one needs to follow. This would instil more confidence when conducting the experiments. A well-sequenced procedure to instruct the students will make the experiment more logical and understandable. During the planning of the procedure, the activity would encourage the students to consider the adequateness and appropriateness of the amounts of chemicals and apparatus used. This would then help to improve the effectiveness and efficiency of data collection.

This book is a continuation of our previous five books — *Understanding Advanced Physical Inorganic Chemistry*, *Understanding Advanced Organic and Analytical Chemistry*, *Understanding Advanced Chemistry Through Problem Solving*, *Understanding Basic Chemistry* and *Understanding Basic Chemistry Through Problem Solving*, retaining the main refutational characteristics of the series by strategically planting think-aloud questions to promote conceptual understanding during experimental planning. It is hoped that these essential questions would make learners aware of the rationale behind each procedural step, the amounts of chemical used and types of apparatus that are appropriate for the experiment.

ACKNOWLEDGEMENTS

We would like to express our sincere thanks to the staff at World Scientific Publishing Co. Pte. Ltd. for the care and attention which they have given to this book, particularly our editors Lim Sook Cheng and Sandhya Devi, our editorial assistant Chow Meng Wai and Stallion Press.

Special appreciation goes to Ms Ek Soo Ben (Principal of Victoria Junior College), Mrs Foo Chui Hoon, Mrs Toh Chin Ling and Mrs Ting Hsiao Shan for their unwavering support to Kim Seng Chan.

Special thanks go to all our students who have made our teaching of chemistry fruitful and interesting. We have learnt a lot from them just as they have learnt some good chemistry from us.

Finally, we thank our families for their wholehearted support and understanding throughout the period of writing this book. We would like to share with all the passionate learners of chemistry two important quotes from the *Analects of Confucius*:

學而時習之，不亦悅乎？(Isn't it a pleasure to learn and practice what is learned time and again?)

學而不思則罔，思而不學則殆 (Learning without thinking leads to confusion, thinking without learning results in wasted effort)

Kim Seng Chan
BSc (Hons), PhD, PDGE (Sec),
MEd, MA (Ed Mgt), MEd (G Ed), MEd (Dev Psy)

Jeanne Tan
BSc (Hons), PDGE (Sec), MEd (LST)

CONTENTS

Preface   v

Acknowledgements   ix

Chapter 1 Planning Using Titration   1

  1.1 Mixture of Potassium Hydrogen Carbonate and Potassium Chloride   2
  1.2 Mixture of Potassium Carbonate and Potassium Hydroxide   8
  1.3 A Sample of Insoluble $CaCO_3$ 14
  1.4 Determine the Relative Atomic Mass of Sodium   19
  1.5 Determine the Identity of an Acid   21
  1.6 Determine the Solubility of $BaCO_3$ Using Acid–Base Titration/Determine the Solubility Constant, K_{sp} of $BaCO_3$ Using Acid–Base Titration   23
  1.7 Determine the Equilibrium Constant for the Formation of Ethyl Ethanoate   27
  1.8 Determine the Acid Dissociation Constant, K_a, of Ethanoic Acid   31
  1.9 Determine the Value of n in $(COOH)_2 \cdot nH_2O$ Using Acid–Base Titration   34
  1.10 Determine the Value of n in $(COOH)_2 \cdot nH_2O$ Using Redox Titration   38
  1.11 Determine the Value of n in $CuSO_4 \cdot nH_2O$ Using Redox Titration   43
  1.12 Determine the Solubility Product Constant (K_{sp}) for $Ba(IO_3)_2$ Using Redox Titration   48
  1.13 Determine the Partition Coefficient of a Weak Acid Through Titration   50

1.14 Safety Precautions for Titration Experiments   53
1.15 Minimizing Experimental Errors or Increasing Reliability   54

Chapter 2 Planning Using Gravimetric Analysis   55

2.1 Mixture of Potassium Hydrogen Carbonate
     and Potassium Chloride   56
2.2 Mixture of Lead Carbonate and Barium Carbonate   62
2.3 A Sample of Impure Sodium Thiosulfate   66
2.4 Determine the Solubility of a Solid/Determine the K_{sp}
     Value of a Partially Soluble Compound   68
2.5 Determine the Concentration of an Acid Through Gravimetry   73
2.6 Safety Precautions for Gravimetry   74
2.7 Minimizing Experimental Errors or Increasing Reliability   75

Chapter 3 Planning Using the Gas Collection Method   77

3.1 Mixture of Potassium Hydrogen Carbonate
     and Potassium Chloride   79
3.2 A Sample of Insoluble $BaCO_3$ 83
3.3 Determination of the Ideal Gas Constant, R, Using $BaCO_3$ 87
3.4 Determine the Identity of an Acid   90
3.5 Determine the Value of n in $(COOH)_2 \cdot nH_2O$ Using
     Gas Collection   92
3.6 Determine the Decomposition Equation of $RbNO_3$ Using
     Gas Collection   95
3.7 Determine the Concentration of H_2O_2 Using Gas Collection   98
3.8 Safety Precautions for Gas Collection Experiments   100
3.9 Minimizing Experimental Errors or Increasing Reliability   100

Chapter 4 Planning for Energetics Experiments   103

4.1 Determine the Enthalpy Change of Solution
     of Sodium Chloride   104
4.2 Determine the Enthalpy Change of Reaction Between
     Aqueous Na_2CO_3 and Aqueous HCl   109
4.3 Determine the Enthalpy Change of Reaction Between
     Solid MgO and Aqueous HCl   114

4.4	Determine the Enthalpy Change of Reaction Between Solid MgO and Aqueous HCl by Plotting a Graph	118
4.5	Determine the Enthalpy Change of Formation of MgO	121
4.6	Determine the Identity of an Acid using Thermometry	122
4.7	Determine the Concentration of an Acid Through Thermometric Titration	126
4.8	Determine the Purity of a Solid Through Thermometric Titration	129
4.9	Determine the Concentration of an Unknown Base Through Thermometry	134
4.10	Determine Whether the Acid Is a Strong or Weak Acid	137
4.11	Determine the Enthalpy Change of Combustion of an Organic Compound	140
4.12	Determine the Heat Capacity of a Calorimeter	145
4.13	Determine the Enthalpy Change of Combustion of Hexane Taking into Consideration the Heat Absorbed by the Calorimeter	148
4.14	Determine the Enthalpy Change of Hydrogenation of Pent-1-yne	150
4.15	Determine How Boiling Point Is Affected by the Presence of Impurities	152
4.16	Safety Precautions for Thermometry Experiments	154
4.17	Minimizing Experimental Errors or Increasing Reliability	155

Chapter 5 Planning for Kinetics Experiments — **157**

5.1	Determine the Rate of Reaction Between HCl and $CaCO_3$ by Gravimetry	162
5.2	Determine the Rate of Decomposition of H_2O_2 Using the Gas Collection Method	165
5.3	Determine the Rate of Decomposition of H_2O_2 Using Titration	169
5.4	Determine the Order of Reaction Using the Initial Rate Method	174
5.5	Determine the Order of Reaction of $Na_2S_2O_8$ and NaI Using the Initial Rate Method	179

5.6 Determine the Activation Energy, E_a, of the Reaction Between $Na_2S_2O_8$ and NaI Using the Initial Rate Method   184
5.7 Determine How the Concentration of an Acid Affects the Rate of Reaction of Magnesium   188
5.8 Determine the Order of Reaction of Iodination of Propanone Using Colorimetry   191
5.9 Safety Precautions for Kinetics Experiments   194
5.10 Minimizing Experimental Errors or Increasing Reliability   195

Chapter 6 Planning for Electrochemical Experiments   197

6.1 Determine the Cell Potential of a Voltaic Cell   200
6.2 Determine the Effect of Concentration on the Cell Potential   204
6.3 Determine Avogadro's Number Through Electrolysis   205
6.4 Design an Experiment for Copper Purification/Design an Experiment to Electroplate an Object   212
6.5 Design an Experiment to Anodize Aluminum   214
6.6 Determine the Charge of an Ion Through Electrolysis   216
6.7 Safety Precautions for Electrolytic Experiments   218
6.8 Minimizing Experimental Errors or Increasing Reliability   219

Chapter 7 Planning for Inorganic Qualitative Analysis   221

7.1 Sequence to Test for an Unknown Gas   224
7.2 Identifying an Anion   224
7.3 Identifying a Cation   227
7.4 Qualitative Analysis with an External Reagent   230
7.5 Self-Contained Qualitative Analysis   234
7.6 Separation of Ions   242
7.7 Safety Precautions for Qualitative Analysis   245

Chapter 8 Planning for Organic Qualitative Analysis   247

8.1 Common Organic Functional Group Tests   247
8.2 Identifying an Organic Functional Group   251
8.3 Safety Precautions for Organic Qualitative Analysis   256

Chapter 9 Planning for Organic Synthesis   **257**

9.1 Starting the Organic Reaction   257
9.2 Separation of Products   259
9.3 Purification of Compounds   271
9.4 Testing for the Purity of a Substance   274
9.5 Safety Precautions for Organic Synthesis   275

Chapter 10 Planning for Spectrophotometric Analysis   **277**

10.1 Determine the Concentration of the $[Ni(H_2O)_6]^{2+}$ Complex   279
10.2 Determine the Formula of the $[Ni(NH_3)_n]^{2+}$ Complex   283

Index   287

CHAPTER 1

PLANNING USING TITRATION

Titration refers to a general class of experiments in which the known property (e.g., concentration) of a solution is used to infer the unknown property of another solution. A typical titration experiment requires the following apparatus:

— Pipette and pipette filler to measure the volume of the analyte or titrand;
— Burette to contain the standard solution or titrant;
— Conical flask to contain the analyte or titrand;
— Volumetric flask to make a standard solution or for carrying out a dilution; and
— An indicator (optional) such as phenolphthalein, methyl orange, starch, etc. to signal the end of a titration.

A titration set-up is shown as follows:

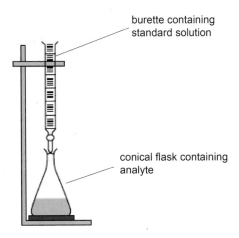

> **Q** What is a standard solution?

A: A standard solution refers to one in which its concentration is already known accurately. Since the concentration of the solution is already known, when we used a certain volume of this standard solution, we would know the number of moles of particles that is used. Hence, we can make use of this information about the number of moles to determine the number of moles of reactants in the analyte, through a balanced chemical equation.

> **Q** Why is a burette used to hold the standard solution?

A: A burette can measure up to two decimal places of accuracy while a pipette can only measure up to one decimal place of accuracy. Since a burette is more accurate than a pipette, it is *usually* used to contain the standard solution. But take note that there are exceptions in which a burette can be used to contain the non-standard solution that is used during titration. For such instances, there are usually other specific reasons for doing it.

1.1 Mixture of Potassium Hydrogen Carbonate and Potassium Chloride

The Task:

A sample of potassium hydrogen carbonate, $KHCO_3$, contains 4–8% of potassium chloride, KCl. You are supposed to plan an experiment using titration to determine the actual percentage of the potassium hydrogen carbonate in the sample with the following chemicals and apparatus:

- 0.100 mol dm^{-3} H_2SO_4 solution;
- 12 g sample of potassium hydrogen carbonate and potassium chloride mixture;
- 250 cm^3 volumetric flask/graduated flask; and
- Standard glassware for titration.

> **Q** Where should we start thinking?

A: (i) What is the purpose of the plan?
— To determine the percentage by mass of the potassium hydrogen carbonate in the mixture.

(ii) What do you need to know in order to determine the mass of potassium hydrogen carbonate?
— We need to know the number of moles of potassium hydrogen carbonate that is present in a fixed mass of the sample that we have measured.

(iii) How do you then determine the number of moles of potassium hydrogen carbonate?
— We can make use of the following balanced equation for the reaction between $KHCO_3$ and H_2SO_4:

$$2KHCO_3(aq) + H_2SO_4(aq) \rightarrow K_2SO_4(aq) + 2CO_2(g) + 2H_2O(l).$$

Thus, if we know the number of moles of H_2SO_4 that was used for the titration, we would be able to deduce the number of moles of $KHCO_3$ that is present in the sample.

(iv) How do you determine the number of moles of H_2SO_4 that is used?
— Since the concentration of the acid is known, from the volume of $H_2SO_4(aq)$ that is used in the titration, we can find out the number of moles of H_2SO_4 used.

(v) Now, what is the expected volume of the acid used during titration?
— We can assume that we have used about 25 cm^3 of aqueous H_2SO_4 to react with 25 cm^3 of the sample $KHCO_3$ solution.

(vi) But, how do you know how much of the solid sample is needed in order to form the solution that is required to react with 25 cm^3 of aqueous H_2SO_4?
— Well, if we know the number of moles of H_2SO_4 used, we would be able to find out the number of moles of $KHCO_3$ in 25 cm^3 of the solution that is pipetted. With this, we would be able to determine the mass of $KHCO_3$ in 25 cm^3 of the solution. Hence, we would be able to find out the mass of the solid sample that we need to measure in order to make 250 cm^3 of the solution.

(vii) Why do you need to have 250 cm^3 of the sample solution?
— To allow us to repeat the titration till we get consistent readings of ±0.10 cm^3.

Pre-Experimental Calculations:

$$2KHCO_3(aq) + H_2SO_4(aq) \rightarrow K_2SO_4(aq) + 2CO_2(g) + 2H_2O(l)$$

Assuming that 25 cm^3 of H$_2$SO$_4$ is used:

Amount of H$_2$SO$_4$ used = $\frac{25}{1000} \times 0.100 = 2.5 \times 10^{-3}$ mol

Assuming that 25 cm^3 of KHCO$_3$ is pipetted:
Amount of KHCO$_3$ present = $2 \times 2.5 \times 10^{-3} = 5.0 \times 10^{-3}$ mol
Molar mass of KHCO$_3$ = 39.1 + 1.0 + 12.0 + 3(16.0) = 100.1 g mol^{-1}
Mass of KHCO$_3$ present = $5.0 \times 10^{-3} \times 100.1 = 0.5005$ g
Assuming that there is 4% by mass of KCl in the sample:

Mass of the impure sample that is needed in order to have 0.5005 g of KHCO$_3$ in 25 cm^3 of the solution = $\frac{100}{96} \times 0.5005 = 0.5213$ g

Hence, mass of the impure sample to be dissolved in 250 cm^3 using the volumetric flask = $0.5213 \times 10 = 5.213$ g.

Q Why is the volume of solution that is pipetted (25 cm^3) for titration about the same as the volume of solution that is used from the burette? Instead, can we pipette 10 cm^3 of solution for the titration?

A: Well, it all depends on the amount of solution that you have for pipetting. If you have only 100 cm^3 of stock solution for pipetting, then it is good to pipette just 10 cm^3 for each titration. This is because if you pipette 25 cm^3 of solution each time, then you can only perform a maximum of three titrations, which is not very ideal as you have a fewer number of trials, which would increase the margin of error. Likewise, if you have 250 cm^3 of stock solution, then although pipetting 10 cm^3 of solution allows you to do a maximum of about 24 titrations, it is unlikely that you will need that many trials. So, it is better to pipette 25 cm^3 rather than 10 cm^3 — why? This is because the error that is incurred in measurement is greater if the amount of substance that is measured is smaller. What do we mean? There is a greater amount of error in the measurement during the measuring of a 10 cm^3 volume than for the 25 cm^3 volume. Assuming that there is an uncertainty in the measurement of the volume, which is about 0.50 cm^3, then the error in the measurement is calculated as shown below:

For 10 cm³, the percentage error in measurement = $\frac{0.50}{10} \times 100\% = 5\%$.

For 25 cm³, the percentage error in measurement = $\frac{0.50}{25} \times 100\% = 2\%$.

Thus, due to the greater amount of error in measuring a smaller quantity of substance, it is better that if we pipette about 25 cm³ of solution, we should also use about 25 cm³ of titrant from the burette. Logically, the error in measurement for a 25–25 cm³ titration is certainly lower than that for a 25–10 cm³ titration. This in turn would be lower than for a 10–10 cm³ titration.

Q Why is it that you can only do a maximum of three titrations if you have 100 cm³ of solution?

A: Before you pipette the solution for analysis, you need to rinse the pipette with the analyte, right? Well, this would "sacrifice" some of the solution for experimental analysis.

Q Why do you assume there is 4% of KCl by mass in the sample? Why do you not assume a value that is higher than 4%?

A: Well, if we assume there is only 4% of KCl, but in reality, there is 8% of KCl, what is the implication? Eight percent of KCl would mean a smaller amount of KHCO₃. Thus, if we measure about 5.213 g of the sample, then we would know for certain that the volume of acid needed to react with the KHCO₃ in the 25 cm³ of solution that has been pipetted is not going to exceed 25 cm³ of the titrant. It is thus smaller than 25 cm³ as there is a smaller amount of KHCO₃ than what we have assumed. But if we have assumed 8% of KCl and in fact there is only 4%, then the volume of the acid that is needed would exceed 25 cm³ of the expected volume based on our theoretical calculations. Basically, our main concern is to ensure that the titration result would not be too high or too low. A very high titration result would make the process tedious and time consuming (as careful refilling of burette is needed for each titration), while a very low titration result would lead to a high percentage error in the titer value (volume of titrant used).

Q Now, since we have found that we actually need about 5 g of the sample, should we use up the entire sample that has been provided at one go?

A: Do not use up the sample at one go. Always leave some sample to allow you to repeat your experiment if there is issue with your first attempt.

 Q Instead of using the volumetric flask, can we use a clean and dry beaker to dissolve the sample in an exact volume of 250 cm^3 of water?

A: No, you can't. When you need to make a standard solution (a solution with concentration that is accurately measured) or doing dilution, you need to use a volumetric flask. Why? This is because even though the volume of the solvent is accurately measured, let's say using a burette, the moment the solute is added into the solvent, the total volume will change due to contraction or expansion during the dissolution or dilution process. A volumetric flask is calibrated with a very precise volume which ensures that the final volume after dilution or dissolution is really that same capacity that has been stated on the flask, as long as the level of the solution reaches the graduation mark.

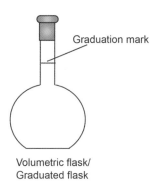

Volumetric flask/ Graduated flask

The Procedure:

The general procedure for titration experiments involves the making of standard solution or carrying out dilution, pipetting of solution, filling up a burette, addition of titrant, etc. The main objective is to determine the volume of a solution with a known concentration that is required to react with a specific volume of an unknown solution that is pipetted. The accuracy of determining this unknown volume is important. Thus, the procedure of a typical titration experiment would consist of a series of steps which inform the student: (1) which step should come first; (2) what apparatus should he/she use; (3) what is the quantity of substance that he/she should measure; and (4) if needed, the reaction conditions such as temperature and pressure.

Procedure for making a standard solution:

(i) *Weigh* accurately *5.2 g* of the sample.
(ii) *Dissolve* the sample in about *50* cm^3 of water in a *beaker*.
(iii) *Transfer* the solution into a *250* cm^3 *volumetric flask* after ensuring that all the solids have dissolved. *Rinse* the beaker with water *a few times* and *transfer the washings* into the volumetric flask, to ensure *quantitative transfer*.
(iv) *Top up* the solution in the volumetric flask to the *graduation mark* using a *dropper*. *Shake* the flask well to get a *homogeneous solution*.

Q What is the meaning of 'quantitative transfer?'

A: Well, it simply means that you have transferred what you have measured quantitatively without loss.

Procedure for titration:

(i) Set up the *50* cm^3 *burette* containing 0.100 mol dm^{-3} H$_2$SO$_4$ solution.
(ii) *Pipette 25.0* cm^3 of the sample solution into a *conical flask*.
(iii) Add two drops of *methyl orange indicator* into the conical flask.
(iv) Take the *initial burette reading*. *Titrate* the sample solution *against* the H$_2$SO$_4$ solution. *Swirl* continuously during the addition of the titrant.
(v) Toward the *end-point*, add the H$_2$SO$_4$ solution *dropwise* and *swirl*. Stop the addition when *one drop of the titrant* causes the indicator to change color from *yellow to orange*.
(vi) Take the *final burette reading* and calculate the *titer volume*.

(vii) *Repeat titrations* until the titer volumes are *within $\pm 0.10\,cm^3$ consistency* (i.e., obtain at least two titer volumes that are within $0.10\,cm^3$ of each other).

(viii) *Repeat the experiment* to check for *reliability of the result*.

> **Q** Can we use phenolphthalein as the indicator for the titration?

A: The working range of the phenolphthalein is from pH 8.3–10, which means that phenolphthalein will only change color when the pH value of the solution hits this range. Coincidentally, at this pH range, HCO_3^- is not converted to CO_2 and H_2O yet. So, you need to use methyl orange as the indicator (working range: pH 3.1–4.4) as by the time the indicator change color at the pH range, all the HCO_3^- would have already been converted to CO_2 and H_2O. For more details, please refer to *Understanding Advanced Physical Inorganic Chemistry* by J. Tan and K. S. Chan.

> **Q** What other reaction mixtures can also make use of the same planning outline as the above?

A: Many! For example: K_2CO_3 with KCl and KOH with KCl.

1.2 Mixture of Potassium Carbonate and Potassium Hydroxide

The Task:

A sample solution containing potassium carbonate, K_2CO_3, is contaminated with KOH. There are 30 g of potassium ions in $1\,dm^3$ of the solution. You are supposed to plan an experiment using titration to determine the actual concentration of the potassium carbonate in the sample with the following chemicals and apparatus:

- $0.100\,mol\,dm^{-3}$ HCl solution;
- $100\,cm^3$ of potassium carbonate and potassium hydroxide solution;
- $250\,cm^3$ volumetric flask/graduated flask; and
- Standard glassware for titration.

> **Q** Where should we start thinking?

A: (i) What is the purpose of the plan?
— To determine the concentration of the potassium carbonate in the mixture.

(ii) What do you need to know in order to determine the concentration of potassium carbonate?
— We need to know the number of moles of potassium carbonate present in a fixed volume of the sample solution that we have measured.

(iii) How do you then determine the number of moles of potassium carbonate?
— We can make use of the following balanced equation for the reaction between K_2CO_3 and HCl:

$$K_2CO_3(aq) + 2HCl(aq) \rightarrow 2KCl(aq) + CO_2(g) + H_2O(l).$$

Thus, if we know the number of moles of HCl that is used for the titration, we would be able to deduce the number of moles of K_2CO_3 present in the sample solution.

(iv) But, wouldn't the KOH in the sample solution also react with the HCl?
— Yes, the KOH would also react with the HCl. So, we need to have an estimation of the amount of HCl that is needed. This can be done because the number of moles of potassium ions that is present is equivalent to the number of moles of HCl that is needed, as from the following two equations:

$$K_2CO_3(aq) + 2HCl(aq) \rightarrow 2KCl(aq) + CO_2(g) + H_2O(l);$$
$$KOH(aq) + HCl(aq) \rightarrow KCl(aq) + H_2O(l).$$

Hence, this is the reason why the question mentioned the concentration of the potassium ions in the solution.

(v) How do you find out the number of moles of HCl that is used?
— From the volume of HCl(aq) that is used in the titration as the concentration of the acid is known.

(vi) Now, what is the expected volume of the HCl acid that is used during titration?
— We can assume that about 25 cm^3 of the HCl is used to react with 25 cm^3 of the sample solution containing K_2CO_3 and KOH.

Q For the total amount of HCl added, how much of it would actually react with the K_2CO_3? How can you find out?

A: The CO_3^{2-} ion would react with the H^+ from HCl in two stages:

$$CO_3^{2-} + H^+ \rightarrow HCO_3^-;$$
$$HCO_3^- + H^+ \rightarrow CO_2 + H_2O.$$

If we pipette 25.0 cm^3 of the mixture, at the point when the HCO_3^- is formed, the pH of the solution is about 9. If we have added phenolphthalein into the solution before we start the titration, at this point, the phenolphthalein would change from pink to colorless. If there are any OH^- ions present in the beginning, together with the CO_3^{2-}, the OH^- would also react with the H^+:

$$OH^- + H^+ \rightarrow H_2O.$$

Hence, the volume of H^+ that is used when the phenolphthalein changes color corresponds to that needed to react with the OH^- and also to convert the CO_3^{2-} to HCO_3^-.

Now, imagine if we pipette another 25.0 cm^3 of the same solution, but with methyl orange as the indicator. When we add H^+ solution to the point when methyl orange changes from yellow to orange, the volume of acid that is used is needed to convert the CO_3^{2-} to CO_2 and H_2O, in addition to reacting with the OH^-. Hence, by taking the difference between these two volumes, we would obtain the volume of acid that is required to convert the HCO_3^- to CO_2 and H_2O. This is known as the double-indicators method.

Pre-Experimental Calculations:

$$K_2CO_3(aq) + 2HCl(aq) \rightarrow 2KCl(aq) + CO_2(g) + H_2O(l);$$
$$KOH(aq) + HCl(aq) \rightarrow KCl(aq) + H_2O(l).$$

Concentration of K^+ in the sample solution = $\frac{30}{39.1}$ = 0.767 mol dm^{-3}

Assuming that 25 cm^3 of mixture of K_2CO_3 and KOH is pipetted,

Amount of K^+ present = $\frac{25}{1000} \times 0.767 = 1.92 \times 10^{-2}$ mol

From the above two equations, 1.92×10^{-2} mol of K^+ would "need" 1.92×10^{-2} mol of HCl.

Therefore, volume of 0.100 mol dm^{-3} HCl needed = $1.92 \times 10^{-2}/0.100$ = 0.192 dm^3 = 192 cm^3.

As the volume of HCl needed is far too much, dilution of the sample solution is needed.

Carrying out the dilution:

Assuming that 25 cm^3 of 0.100 mol dm^{-3} HCl is used to react with 25 cm^3 of the sample solution:

Amount of HCl used = $\frac{25}{1000} \times 0.100 = 2.5 \times 10^{-3}$ mol

Amount of K$^+$ present = Amount of HCl used = 2.5×10^{-3} mol

Therefore, concentration of K$^+$ = $\frac{2.5 \times 10^{-3}}{25/1000}$ = 0.100 mol dm^{-3}.

If we need to prepare a 250 cm^3 of 0.100 mol dm^{-3} K$^+$ solution from the 0.767 mol dm^{-3} sample solution, the volume of 0.767 mol dm^{-3} sample solution that is needed (V dm^3) is calculated as shown:

$0.250 \times 0.100 = V \times 0.767$ {*based on the concept that the number of moles do not change during dilution*}

$\Rightarrow$ V = 0.0326 dm^3 = 32.6 cm^3.

Hence, we need to dilute 32.0 cm^3 of 0.767 mol dm^{-3} sample solution to form a 250 cm^3 solution with a concentration of approximately 0.100 mol dm^{-3}.

So, in order to get an idea of the volume, concentration or the mass of reagents that are needed, we need to start off the pre-calculations by making some assumptions?

A: You are right!

The Procedure:

Procedure for dilution:

(i) Introduce *32.60* cm^3 of the sample solution into a *250* cm^3 *volumetric flask* using a *50* cm^3 *burette*.

(ii) *Top up* the solution in the volumetric flask to the *graduation mark* using a *dropper*. Shake the flask well to get a *homogeneous solution*.

Procedure for titration:
  (i) Set-up the *50 cm^3 burette* containing 0.100 mol dm^{-3} *HCl solution*.
  (ii) *Pipette 25.0 cm^3* of the diluted sample solution into a *conical flask*.
  (iii) Add two drops of *phenolphthalein indicator* into the conical flask.
  (iv) Take the *initial burette reading*. Titrate the sample solution *against the HCl solution*. Swirl continuously during the addition of the titrant.
  (v) Toward the *end-point*, add the HCl solution *dropwise* and *swirl*. Stop the addition when *one drop* of the titrant causes the indicator to change from *pink to colorless*.
  (vi) Take the *final burette reading* and calculate the *titer volume*.
  (vii) *Repeat titrations* until the titer volumes are within *±0.10 cm^3 consistency* (i.e., get at least two titer volumes that are within 0.10 cm^3 of each other).
  (viii) *Pipette another 25.0 cm^3* of the diluted sample solution into a *conical flask*.
  (ix) Add two drops of *methyl orange indicator* into the *conical flask*.
  (x) Take the *initial burette reading*. Titrate the sample solution *against the HCl solution*. Swirl continuously during the addition of the titrant.
  (xi) Toward the *end-point*, add the HCl solution *dropwise* and *swirl*. Stop the addition when one drop of the titrant causes the indicator to change from *yellow to orange*.
  (xii) Take the *final burette reading* and calculate the *titer volume*.
  (xiii) *Repeat titrations* until the titer volumes are within *±0.10 cm^3 consistency* (i.e., get at least two titer volumes that are within 0.10 cm^3 of each other).

Q Can we immediately add methyl orange to the same solution after the titration using phenolphthalein and then continue the titration from there?

A: Yes, you may. Just take note of the following differences:
Let the volume of HCl that is needed for the following reactions be:

$$CO_3^{2-} + H^+ \rightarrow HCO_3^- \quad x \text{ cm}^3;$$
$$HCO_3^- + H^+ \rightarrow CO_2 + H_2O \quad x \text{ cm}^3; \text{ and}$$
$$OH^- + H^+ \rightarrow H_2O \quad y \text{ cm}^3.$$

Then, for the titration using phenolphthalein as indicator *only*, the volume of HCl used is $(x + y)$ cm^3.

For the titration using methyl orange as indicator *only*, the volume of HCl used is $(2x + y)$ cm^3.

For the titration using phenolphthalein indicator and then *immediately followed by* the addition of methyl orange, the volume of HCl used would be $(x + y)$ cm^3 and x cm^3, respectively.

Q What other reaction mixtures could also make use of the same double-indicators method?

A: The following table shows the possible reaction mixtures:

Reaction Mixture	Using Phenolphthalein Only	Using Methyl Orange Only	Using Phenolphthalein, Immediately Followed by Methyl Orange
HCO_3^- and CO_3^{2-}	$CO_3^{2-} + H^+ \rightarrow HCO_3^-$	$CO_3^{2-} + 2H^+ \rightarrow CO_2 + H_2O$ $HCO_3^- + H^+ \rightarrow CO_2 + H_2O$	$CO_3^{2-} + H^+ \rightarrow HCO_3^-$ then $HCO_3^- + H^+ \rightarrow CO_2 + H_2O$, where the HCO_3^- originates from both the CO_3^{2-} and those that are already present since the beginning.

(Continued)

(Continued)

Reaction Mixture	Using Phenolphthalein Only	Using Methyl Orange Only	Using Phenolphthalein, Immediately Followed by Methyl Orange
HCO_3^- and OH^-	$OH^- + H^+ \rightarrow H_2O$	$HCO_3^- + H^+ \rightarrow CO_2 + H_2O$ $OH^- + H^+ \rightarrow H_2O$	$OH^- + H^+ \rightarrow H_2O$ then $HCO_3^- + H^+ \rightarrow CO_2 + H_2O$
OH^- and NH_3	$OH^- + H^+ \rightarrow H_2O$	$OH^- + H^+ \rightarrow H_2O$ $NH_3 + H^+ \rightarrow NH_4^+$	$OH^- + H^+ \rightarrow H_2O$ then $NH_3 + H^+ \rightarrow NH_4^+$

You can also use the same planning outline for a mixture of two weak organic acids, a mixture of a strong and weak acids or a mixture of two weak organic bases which have different end-point pH values. There are simply too many possibilities to consider, so all you need to know is the concept of double-indicators method and *why* it can be used as such.

1.3 A Sample of Insoluble $CaCO_3$

The Task:

During a field trip, a student obtained some impure limestone, which is an impure form of calcium carbonate, $CaCO_3$. Describe how you would help the student to determine the percentage of $CaCO_3$ in a sample of the impure limestone by a method using titration. Take note that the limestone does not dissolve in water and you are provided with the following chemicals and apparatus:

- 1.50 mol dm^{-3} HCl solution;
- 0.100 mol dm^{-3} NaOH solution;
- 5 g of impure $CaCO_3$;
- 250 cm^3 volumetric flask/graduated flask; and
- Standard glassware for titration.

> **Q** Where should we start thinking?

A: (i) What is the purpose of the plan?
— To determine the percentage of $CaCO_3$ in a sample of impure limestone.

(ii) What do you need to know in order to determine the percentage of $CaCO_3$?
— We need to know the number of moles of $CaCO_3$ that is present in a fixed mass of the sample that we have measured.

(iii) How do you then determine the number of moles of $CaCO_3$?
— We can make use of the following balanced equation for the reaction between $CaCO_3$ and HCl:

$$CaCO_3(s) + 2HCl(aq) \rightarrow CaCl_2(aq) + CO_2(g) + H_2O(l).$$

Thus, if we know the number of moles of HCl that is used for the titration, we would be able to deduce the number of moles of $CaCO_3$ that is present in the sample mass.

(iv) But can you titrate the solid $CaCO_3$ directly?
— No, we need to dissolve the solid $CaCO_3$ in water. But wait a minute, $CaCO_3$ is insoluble in water! So, we need to dissolve the $CaCO_3$ in a known amount of aqueous HCl and then later on, determine the amount of the unreacted "leftover" HCl.

(v) Yes, this is known as back-titration. But how do you find out the number of moles of HCl to be used?
— Well, we can assume a certain mass of $CaCO_3$ is used and then calculate the theoretical amount of aqueous HCl that is needed.

Pre-Experimental Calculations:

$$CaCO_3(s) + 2HCl(aq) \rightarrow CaCl_2(aq) + CO_2(g) + H_2O(l)$$

Assuming that we used about 2.5 g of impure $CaCO_3$ and all the 2.5 g is due to $CaCO_3$ only:
Molar mass of $CaCO_3 = 40.1 + 12.0 + 3(16.0) = 100.1$ g mol^{-1}
Amount of $CaCO_3$ in 2.5 g $= \frac{2.5}{100.1} = 2.50 \times 10^{-2}$ mol
Amount of HCl needed $= 2 \times$ Amount of $CaCO_3$ in 2.5 g
$= 2 \times 2.50 \times 10^{-2} = 5.00 \times 10^{-2}$ mol

Volume of 1.50 mol dm^{-3} HCl solution = $\dfrac{5.00 \times 10^{-2}}{1.5}$ = 0.0333 dm^3
= 33.3 cm^3.

Assuming that we have added 35 cm^3 of the 1.50 mol dm^{-3} HCl solution to dissolve all the CaCO$_3$:

Amount of HCl in 35 cm^3 of the 1.50 mol dm^{-3} HCl solution = $\dfrac{35}{1000} \times 1.5$
= 0.0525 mol

Amount of unreacted HCl left over = 0.0525 − 5.00 × 10^{-2}
= 2.50 × 10^{-3} mol

Volume of 0.100 mol dm^{-3} NaOH solution needed to react with the unreacted HCl = $\dfrac{2.50 \times 10^{-3}}{0.1}$ = 0.025 dm^3 = 25 cm^3.

> **Q** What would happen if the concentration of the NaOH solution that is given is 0.010 mol dm^{-3}?

A: Then, the volume of 0.010 mol dm^{-3} NaOH solution that is needed to react with the unreacted HCl = $\dfrac{2.50 \times 10^{-3}}{0.01}$ = 0.25 dm^3 = 250 cm^3. Thereafter, you need to carry out a dilution of the leftover HCl solution first before titration. So, take note that the question may be set in such a way that in principle, the thinking process to solve the question is very similar to what we have been discussing here. The difference is that there is an additional dilution step that you need to take care of.

> **Q** Are there any other assumptions being made other than assuming that there is only CaCO$_3$ present in the impure sample?

A: Well, we have to also assume that the impurities that are present are soluble in water and that they do not react with the acid used.

> **Q** What happens if there are insoluble impurities?

A: A filtration step would be needed to be included before the titration.

The Procedure:

Procedure for dissolving the insoluble CaCO$_3$:

(i) *Weigh* accurately 2.5 g of the impure limestone in a *weighing bottle*.

(ii) Introduce *35.00* cm³ of the 1.50 mol dm⁻³ *HCl solution* into a *250* cm³ *conical flask* using a *burette*.
(iii) *Carefully*, add the limestone sample from the weighing bottle into the conical flask containing the acid and *immediately cover the conical flask with a filter funnel* to *prevent acid spray* from escaping (as shown in the diagram below).

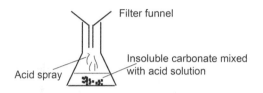

(iv) *Reweigh* the mass of the *emptied weighing bottle. Calculate* the *actual mass* of the solid that has been introduced into the conical flask.
(v) *Rinse* the *acid spray underneath the filter funnel* with deionized water and let the *washing go into the conical flask*.

Q Can we add the acid to the solid carbonate instead?

A: Well, if the acid is "streaming" into the conical flask from the burette, the moment the acid comes in contact with the carbonate, the effervescence that is produced is going to "rush" through the streaming acid. This would cause more acid spray. So, this is not the ideal way of doing the experiment. The good thing about adding the solid carbonate to the acid is that the solid can be added into the conical flask in the shortest possible time.

Procedure for titration:

(i) Set up the 50 cm³ *burette* containing 0.100 mol dm⁻³ *NaOH solution*.
(ii) Add two drops of *phenolphthalein indicator* into the conical flask from above.
(iii) Take the *initial burette reading. Titrate* the sample solution *against the NaOH solution. Swirl* continuously during the addition.
(iv) Toward the *end-point*, add the NaOH solution *dropwise* and *swirl*. Stop the addition when one drop of the titrant causes the indicator to change from *colorless to pink*.
(v) Take the *final burette reading* and calculate the *titer volume*.
(vi) *Repeat* the experiment with another 2.5 g of the sample.

Q So, we can't repeat the titration experiment here?

A: You can't because each weighing of the sample allows you to perform only one titration. But if you have a dilution process, then you should perform multiple titrations until you get consistent results.

Q How would the procedure be like if there is a dilution process involved?

A: Well, it would be as follows:

Procedure for dissolving the insoluble $CaCO_3$:

(i) *Weigh* accurately *XXX* g of the impure limestone in a *weighing bottle*.
(ii) *Introduce XXX* cm^3 of the XXX mol dm^{-3} *HCl solution* into a 250 cm^3 *conical flask using a burette*.
(iii) *Carefully*, add the limestone sample from the weighing bottle into the conical flask containing the acid and *immediately cover the conical flask* with a *filter funnel* to *prevent acid spray* from escaping.
(iv) *Reweigh* the mass of the *emptied weighing bottle*. Calculate the *actual mass* of the solid that has been introduced into the conical flask.
(v) *Rinse* the *acid spray underneath the filter funnel* with deionized water and let the *washing go into the conical flask*.
(vi) *Transfer* the solution into the *volumetric flask* after ensuring that *all the solids have dissolved*. Rinse the conical flask with water *a few times* and *transfer the washings* into the volumetric flask, to ensure *quantitative transfer*.
(vii) *Top up* the solution in the volumetric flask to the *graduation mark* using a *dropper*. Shake the flask well to get a *homogeneous solution*.

Procedure for titration:

(i) Set-up the 50 cm^3 *burette* containing 0.100 mol dm^{-3} *NaOH solution*.
(ii) *Pipette* 25.0 cm^3 of acid solution into a *conical flask*.
(iii) Add two drops of *phenolphthalein indicator* into the conical flask.
(iv) Take the *initial burette reading*. *Titrate* the sample solution *against the NaOH solution*. *Swirl* continuously during the addition.

(v) Toward the *end-point*, add the NaOH solution *dropwise* and *swirl*. Stop the addition when *one drop of the titrant* causes the indicator to change from *colorless to pink*.
(vi) Take the *final burette reading* and calculate the *titer volume*.
(vii) *Repeat* titrations until the titer volumes are within ±*0.10* cm^3 *consistency* (i.e., get at least two titer volumes that are within 0.10 cm^3 of each other).

Note: In addition, such a planning process can also be used for the following similar cases:

(i) Determine the relative atomic mass of the element **M** in the metal carbonate, MCO_3.
(ii) Determine the concentration of a H_2O_2 solution through redox titration using acidified $KMnO_4$.

The thinking processes involved in these tasks would be very similar to what we have discussed in Section 1.3. The main idea is to obtain the number of moles of MCO_3 or H_2O_2 indirectly via the titration method with the help of a balanced chemical equation.

1.4 Determine the Relative Atomic Mass of Sodium

The Task:

You are given a piece of sodium metal. Describe how you would determine the relative atomic mass of the sodium using simple titration. You are provided with the following chemicals and apparatus:

- 1.50 mol dm^{-3} HCl solution;
- 0.100 mol dm^{-3} NaOH solution;
- 5 g of sodium metal;
- 250 cm^3 volumetric flask/graduated flask; and
- Standard glassware for titration.

> **Q** Where should we start thinking?

A: (i) What is the purpose of the plan?
— To determine the relative atomic mass of the sodium metal.

(ii) What do you need to know in order to determine the relative atomic mass of the sodium?
— We need to know the number of moles of Na in a fixed mass of the sample that we have measured since the amount in moles of Na is calculated as follows:

$$\text{Amount of Na in moles} = \frac{\text{Mass of Na used}}{\text{Relative atomic mass of Na}}$$

Hence, if both the amount of Na in moles and the mass of Na used are known, the value of the relative atomic mass of Na can be easily calculated.

(iii) How do you then determine the number of moles of the sodium?
— We can make use of the following balanced equation between the reaction of Na and HCl:

$$2Na(s) + 2HCl(aq) \rightarrow 2NaCl(aq) + H_2(g).$$

Thus, if we know the number of moles of HCl that is used for the titration, we would be able to deduce the number of moles of Na that is present in the sample that we have weighed.

(iv) But can you titrate the solid Na directly?
— No, we need to dissolve the Na metal in a known amount of HCl solution and then determine the amount of the unreacted "leftover" HCl.

(v) Yes, this is known as back-titration. But how do you find out the number of moles of HCl to be used?
— Well, we can assume that a certain mass of Na metal that is to be used and then calculate the theoretical amount of HCl solution that is needed.

> **Q** But the reaction of sodium with HCl is too vigorous. Is there a better method?

A: Well, if the question does not provide you with aqueous HCl solution, what you can do is to simply react the sodium with water:

$$Na(s) + H_2O(l) \rightarrow NaOH(aq) + \tfrac{1}{2} H_2(g).$$

Then, titrate the NaOH that is formed with the given HCl solution. From the number of moles of HCl used, you can determine the number of moles of sodium from the equation below:

$$NaOH(aq) + HCl(aq) \rightarrow NaCl(aq) + H_2O(l).$$

So, do you see the similarities between Sections 1.3 and 1.4? Both employ the back titration method. The pre-experimental calculations which would lead to the procedure for titration are very similar to that discussed in Section 1.3.

1.5 Determine the Identity of an Acid

The Task:

You are given an acid of concentration, 9 g dm^{-3}. This acid can be one of the following monobasic acids:

- HA of M_r 42.9;
- HB of M_r 57.3; or
- HC of M_r 90.0.

Describe how you would determine the identity of the acid using simple titration. You are provided with the following chemicals and apparatus:

- 10 g of solid sodium hydroxide, NaOH;
- 250 cm^3 volumetric flask/graduated flask; and
- Standard glassware for titration.

Q Where should we start thinking?

A: (i) What is the purpose of the plan?
— To determine the identity of the acid.

(ii) What do we need to know in order to determine the identity of the acid?
— Since the concentration (9 g dm^{-3}) of the acid is known, we can convert the mass concentration into molar concentration in mol dm^{-3}. This would mean that there are three possible concentrations depending on which acid it is:

- HA of concentration = $\frac{9}{42.9}$ = 0.210 mol dm^{-3};
- HB of concentration = $\frac{9}{57.3}$ = 0.157 mol dm^{-3}; or
- HC of concentration = $\frac{9}{90.0}$ = 0.100 mol dm^{-3}.

If we know the number of moles of NaOH required to react with 25 cm^3 of the acid solution, we would then be able to calculate the concentration of the acid. Hence, once we know the concentration of the acid, its identity will be known.

(iii) How do you then determine the number of moles of NaOH that is used?

— We can make use of the following balanced equation for the reaction between NaOH and HA/HB/HC:

$$NaOH(aq) + HA/HB/HC(aq) \rightarrow NaA/NaB/NaC(aq) + H_2O(l).$$

Thus, if we know the number of moles of NaOH used for the titration, we would be able to deduce the number of moles of HA/HB/HC present in a particular volume of the sample solution that we have used.

(iv) But you are given solid sodium hydroxide, so how are you going to use it?

— We need to make a standard solution of NaOH first and then use it for the titration.

(v) What is the concentration of the standard NaOH solution that you should make?

— If we take 25 cm^3 of the acid to react with 25 cm^3 of the NaOH solution, we would be able to calculate a range of concentrations of NaOH solution needed to react with 25 cm^3 of the acid.

Pre-Experimental Calculations:

$$NaOH(aq) + HA/HB/HC(aq) \rightarrow NaA/NaB/NaC(aq) + H_2O(l).$$

Assuming that 25 cm^3 of HA solution is used:

Amount of HA used = $\frac{25}{1000} \times 0.210 = 5.25 \times 10^{-3}$ mol

Amount of NaOH needed = Amount of HA used = 5.25×10^{-3} mol

Concentration of NaOH = $\frac{5.25 \times 10^{-3}}{25/1000} = 0.21$ mol dm^{-3}

Assuming that 25 cm^3 of HB is used:

Amount of HB used = $\frac{25}{1000} \times 0.157 = 3.93 \times 10^{-3}$ mol

Amount of NaOH needed = Amount of HB used = 3.93×10^{-3} mol
Concentration of NaOH = $\dfrac{3.93 \times 10^{-3}}{25/1000}$ = 0.157 mol dm^{-3}
Assuming that 25 cm^3 of HC is used:
Amount of HC used = $\dfrac{25}{1000} \times 0.100 = 2.5 \times 10^{-3}$ mol
Amount of NaOH needed = Amount of HC used = 2.5×10^{-3} mol
Concentration of NaOH = $\dfrac{2.5 \times 10^{-3}}{25/1000}$ = 0.100 mol dm^{-3}.
Hence, the concentration of the standard NaOH solution can range from 0.100–0.21 mol dm^{-3}. We would use the value of 0.15 mol dm^{-3} for the concentration of the standard NaOH solution.
Molar mass of NaOH = 23.0 + 16.0 + 1.0 = 40.0 g mol^{-1}
Mass of solid NaOH that is needed to make 250 cm^3 NaOH solution
$$= \tfrac{1}{4} \times 0.15 \times 40.0 = 1.5 \text{ g}.$$
The procedure for the preparation of the standard solution and titration would be similar to what we have discussed before (see Section 1.1).

1.6 Determine the Solubility of BaCO$_3$ Using Acid–Base Titration/Determine the Solubility Constant, K_{sp} of BaCO$_3$ Using Acid–Base Titration

The Task:

You are given some pure barium carbonate, BaCO$_3$. Plan a titration experiment to determine whether the presence of carbon dioxide in water does increase the solubility of BaCO$_3$. You are provided with the following chemicals and apparatus:

- Solid BaCO$_3$;
- Standard HCl solution;
- Deionized water;
- Deionized water saturated with carbon dioxide;
- Phenolphthalein indicator;
- Methyl orange indicator; and
- Standard glassware for titration.

> **Q** Where should we start thinking?

A: (i) What is the purpose of the plan?
— To determine whether the presence of CO_2 increases the solubility of $BaCO_3$.

(ii) What do you need to do then?
— We need to dissolve some $BaCO_3$ in pure water and some $BaCO_3$ in water that has already been saturated with CO_2 gas. Then, determine the amount of $BaCO_3$ that has dissolved in each of the two solutions.

(iii) What do you need to know in order to determine how much $BaCO_3$ has dissolved?
— We can filter the mixture first, then pipette the filtrate and titrate the filtrate with the given standard HCl solution.

(iv) How do you then determine the number of moles of $BaCO_3$?
— We can make use of the following balanced equation for the reaction between $BaCO_3$ and HCl:

$$2H^+(aq) + CO_3^{2-}(aq) \rightarrow CO_2(g) + H_2O(l).$$

Thus, if we know the number of moles of HCl that is used for the titration, we would be able to deduce the number of moles of dissolved $BaCO_3$ present in the solution. Next, we can then calculate the mass of dissolved $BaCO_3$.

> **Q** Can we do a pre-calculation to find out how much HCl solution is needed for the titration?

A: No and yes! No because you do not know the solubility of the $BaCO_3$, thus you can't do a pre-calculation. Yes, provided the question has given you an estimated solubility value. Anyway, the whole exercise here is to let you see how the thinking process behind the planning exercise is like.

The Procedure:

Procedure for dissolving the insoluble $BaCO_3$:

(i) Use a 50 cm³ *burette* to introduce *100 cm³* of *deionized water* into a *clean and dry conical flask*.

(ii) Introduce some solid barium carbonate and *stir* the solution until *some solid remains undissolved*.
(iii) *Filter* the mixture into a *clean and dry conical flask*.
(iv) *Repeat* steps (i) to (iii) using *deionized water that has been saturated with CO_2*.

> **Q** Why do we need "clean and dry conical flask?"

A: If the conical flask is wet, the extra water present would cause more $BaCO_3$ to dissolve. This would defeat the purpose of using an accurate burette to measure the volume of the deionized water. In addition, if the filtrate comes in contact with "extra water," then the concentration of the $BaCO_3$ that we are going to determine later would not be accurate anymore.

Procedure for titration:

(i) Set up the 50 cm^3 *burette* containing the standard *HCl solution*.
(ii) *Pipette* 25.0 cm^3 of the filtrate into a *conical flask*.
(iii) Add two drops of *methyl orange indicator* into the conical flask.
(iv) Take the *initial burette reading*. *Titrate* the sample solution *against the HCl solution*. *Swirl* continuously during the addition of the titrant.
(v) Toward the *end-point*, add the HCl solution *dropwise* and *swirl*. Stop the addition of the titrant when *one drop of the titrant* causes the indicator to change from *yellow to orange*.
(vi) Take the *final burette reading* and calculate the *titer volume*.
(vii) *Repeat* titrations until the titer volumes are within ±0.10 cm^3 *consistency* (i.e., get at least two titer volumes that are within 0.10 cm^3 of each other).
(viii) *Repeat steps (ii) to (vii)* for the filtrate of which $BaCO_3$ is dissolved in deionized water that has been saturated with CO_2.
(ix) *Repeat the experiment* to check for *reliability of results*.

> **Q** Can we use phenolphthalein as the indicator?

A: Well, this is a weak base–strong acid titration, which means that the pH at the end-point is in the acidic range. Methyl orange as an indicator would be more suitable as its working range coincides with the drastic change of pH at the end-point for such a weak base–strong acid titration.

> **Q** After the experiment, how can we make use of the solubility data for the $BaCO_3$?

A: Good question! You can use it to calculate the solubility product constant for $BaCO_3$:

$$BaCO_3(s) \rightleftharpoons Ba^{2+}(aq) + CO_3^{2-}(aq), \quad K_{sp} = [Ba^{2+}(aq)][CO_3^{2-}(aq)].$$

If the solubility of $BaCO_3$ is x mol dm^{-3}, then $K_{sp} = [Ba^{2+}(aq)][CO_3^{2-}(aq)] = x^2$ mol^2 dm^{-6}.

Note: In addition, such a planning process can also be used for the following similar cases:

(i) Determine the K_{sp} of $Pb(OH)_2$:

$$Pb(OH)_2(s) \rightleftharpoons Pb^{2+}(aq) + 2OH^-(aq), \quad K_{sp} = [Pb^{2+}(aq)][OH^-(aq)]^2.$$

(ii) Determine the K_{sp} of $Cu(OH)_3$ in the presence of aqueous NaOH:

$$Cu(OH)_2(s) \rightleftharpoons Cu^{2+}(aq) + 2OH^-(aq), \quad K_{sp} = [Cu^{2+}(aq)][OH^-(aq)]^2.$$

(Note: Firstly, the concentration of the aqueous NaOH (x mol dm^{-3}) solution must be determined through titration. Then, a certain amount of solid $Cu(OH)_2$ is dissolved in a fixed volume of the NaOH solution. This solution is then filtered and titrated. The difference in the concentration of the OH$^-$ ions would give us the concentration of the Cu^{2+} ion (y mol dm^{-3}):

$$Cu(OH)_2(s) \rightleftharpoons Cu^{2+}(aq) + 2OH^-(aq).$$

Initial conc./mol dm^{-3}	0	x
At equilibrium/mol dm^{-3}	y	$(x+2y)$

$K_{sp} = [Cu^{2+}(aq)][OH^-(aq)]^2 = y(x+2y)^2$ mol^3 dm^{-9}).

Q So, does it mean that the ratio of $[Cu^{2+}(aq)] : [OH^-(aq)] \neq 1:2$ when we dissolve solid $Cu(OH)_2$ in a solution that has already contained some OH^- ions?

A: Yes, indeed. Due to the common ion effect, the solubility of the solid $Cu(OH)_2$ is suppressed. Hence, the ratio of $[Cu^{2+}(aq)] : [OH^-(aq)] \neq 1:2$; it would have been 1:2 if the solid $Cu(OH)_2$ is dissolved in pure water. So, take note that the question can be modified to ask you to calculate the K_{sp} of $Cu(OH)_2$ in a solution containing Cu^{2+} ions. Then, everything is similar.

1.7 Determine the Equilibrium Constant for the Formation of Ethyl Ethanoate

The Task:

Ethanol reacts with ethanoic acid to give the ester, ethyl ethanoate:

$$CH_3CH_2OH(aq) + CH_3COOH(aq) \rightleftharpoons CH_3COOCH_2CH_3(aq) + H_2O(l).$$

The equilibrium constant for the above reaction is defined as,

$$K_C = \frac{[CH_3COOCH_2CH_3]}{[CH_3CH_2OH][CH_3COOH]}.$$

Describe how you would determine the equilibrium constant using simple titration. You are provided with the following chemicals and apparatus:

- 100 cm^3 of 1.00 mol dm^{-3} ethanol solution;
- 100 cm^3 of 1.00 mol dm^{-3} ethanoic acid solution;
- 100 cm^3 of 0.500 mol dm^{-3} NaOH solution;
- 250 cm^3 volumetric flask/graduated flask; and
- Standard glassware for titration.

Q Where should we start thinking?

A: (i) What is the purpose of the plan?
— To determine the equilibrium constant.

(ii) What do you need to know in order to determine the equilibrium constant?
— We need to determine the concentration of the ethanoic acid at equilibrium. But in order to do this, we need to know the number of moles of CH_3COOH present in a fixed volume of the sample solution that we have measured.

(iii) How do you then determine the number of moles of CH_3COOH?
— We can make use of the following balanced equation for the reaction between CH_3COOH and NaOH:

$$NaOH(aq) + CH_3COOH(aq) \rightarrow CH_3COONa(aq) + H_2O(l).$$

Thus, if we know the number of moles of NaOH used for the titration, we would be able to deduce the number of moles of CH_3COOH present in the sample solution.

(iv) Now, what is the expected volume of the NaOH solution to be used during the titration?
— Well, we can assume that about 25 cm^3 of the NaOH solution is used to react with 25 cm^3 of the CH_3COOH in the solution mixture.

Pre-Experimental Calculations:

$$CH_3CH_2OH(aq) + CH_3COOH(aq) \rightleftharpoons CH_3COOCH_2CH_3(aq) + H_2O(l)$$

Assuming that 50 cm^3 of 1.00 mol dm^{-3} ethanol is mixed with 50 cm^3 of 1.00 mol dm^{-3} ethanoic acid solution:

Concentration of $CH_3COOH(aq) = \frac{1}{2} \times 1.00 = 0.50$ mol dm^{-3}

Assuming that none of the ethanoic acid has reacted:

Amount of $CH_3COOH(aq)$ in 25.0 cm^3 mixture = $\frac{25}{1000} \times 0.50$

$= 1.25 \times 10^{-2}$ mol

$$NaOH(aq) + CH_3COOH(aq) \rightarrow CH_3COONa(aq) + H_2O(l)$$

Amount of NaOH needed = Amount of $CH_3COOH(aq) = 1.25 \times 10^{-2}$ mol

Volume of 0.500 mol dm^{-3} NaOH solution needed = $\frac{1.25 \times 10^{-2}}{0.50}$
$= 2.5 \times 10^{-2}$ dm^3 = 25 cm^3.

The Procedure:

Procedure for making the equilibrium mixture:

(i) Use a 50 cm^3 *burette* to introduce *50.00* cm^3 of 1.00 mol dm^{-3} ethanol solution into a *clean and dry beaker*.
(ii) Use another 50 cm^3 *burette* to introduce *50.00* cm^3 of 1.00 mol dm^{-3} ethanoic acid solution into the same beaker.
(iii) *Stir* the solution and *cover it to prevent evaporation*. Let the *mixture stand for a day*.

Procedure for titration:

(i) Set up the 50 cm^3 *burette* containing 0.500 mol dm^{-3} *NaOH solution*.
(ii) *Pipette 25.0* cm^3 of the sample solution into a *conical flask*.
(iii) Add two drops of *phenolphthalein indicator* into the conical flask.
(iv) Take the *initial burette reading. Titrate* the sample solution *against the NaOH solution. Swirl* continuously during the addition of the titrant.
(v) Toward the *end-point*, add the NaOH solution *dropwise* and *swirl*. Stop the addition when *one drop of titrant* causes the indicator to change from *colorless to pink*. The titration should be completed in the shortest possible time.
(vi) Take the *final burette reading* and calculate the *titre volume*.
(vii) *Repeat titrations* until the titer volumes are within ±*0.10* cm^3 *consistency* (i.e., get at least two titer volumes that are within 0.10 cm^3 of each other).
(viii) *Repeat the experiment* to check for *reliability of results*.

Treatment of Results:

Let the [CH$_3$COOH] at equilibrium that is determined from the titration be x mol dm^{-3}.

$$CH_3CH_2OH(aq) + CH_3COOH(aq) \rightleftharpoons CH_3COOCH_2CH_3(aq) + H_2O(l)$$

Initial conc./ 0.50   0.50   0
mol dm^{-3}

At equilibrium x x $(0.50 - x)$

$$K_C = \frac{[CH_3COOCH_2CH_3]}{[CH_3CH_2OH][CH_3COOH]} = \frac{(0.50-x)}{(x)^2} \text{ mol}^{-1} \text{ dm}^3.$$

> **Q** Why isn't the [H$_2$O] included in the equilibrium expression,
> $$K_C = \frac{[CH_3COOCH_2CH_3]}{[CH_3CH_2OH][CH_3COOH]}?$$

A: As water is the medium and present in large excess, the [H$_2$O] is a constant. So, you can perceive the equilibrium expression,

$$K_C = \frac{[CH_3COOCH_2CH_3]}{[CH_3CH_2OH][CH_3COOH]},$$ as a "mutated" equilibrium constant from

this one: $K_C = \dfrac{[CH_3COOCH_2CH_3][H_2O]}{[CH_3CH_2OH][CH_3COOH]}$. This means that, if at the start you have used pure ethanoic acid and pure ethanol to establish the equilibrium system, then the equilibrium expression should be as follows:

$$CH_3CH_2OH(l) + CH_3COOH(l) \rightleftharpoons CH_3COOCH_2CH_3(l) + H_2O(l),$$

$$K_C = \frac{[CH_3COOCH_2CH_3][H_2O]}{[CH_3CH_2OH][CH_3COOH]}.$$

The calculation would be as follows:
Let the initial [CH$_3$COOH] and [CH$_3$CH$_2$OH] after mixing be y mol dm^{-3}.
Let the [CH$_3$COOH] at equilibrium that is determined from the titration be x mol dm^{-3}.

	CH$_3$CH$_2$OH(l)	+ CH$_3$COOH(l)	$\rightleftharpoons$ CH$_3$COOCH$_2$CH$_3$(l)	+ H$_2$O(l)
Initial conc./ mol dm^{-3}	y	y	0	0
At equilibrium/ mol dm^{-3}	x	x	$(y-x)$	$(y-x)$

$$K_C = \frac{[CH_3COOCH_2CH_3][H_2O]}{[CH_3CH_2OH][CH_3COOH]} = \frac{(y-x)^2}{(x)^2}.$$

> **Q** As the ethanoic acid is titrated with the aqueous NaOH, wouldn't the decreasing concentration of the ethanoic acid causes the position of equilibrium to shift left, hence affect the accuracy of the result?

A: You are right to note that the position of equilibrium would in fact shift to the left. That is why you need to finish the titration in the shortest possible time.

1.8 Determine the Acid Dissociation Constant, K_a, of Ethanoic Acid

The Task:

Ethanoic acid is a weak acid which does not fully dissociate in water:

$$CH_3COOH(aq) + H_2O(l) \rightleftharpoons CH_3COO^-(aq) + H_3O^+(aq).$$

The acid dissociation constant for above reaction is defined as

$$K_a = \frac{[CH_3COO^-][H_3O^+]}{[CH_3COOH]}.$$

The acid dissociation constant, K_a, of a weak acid can be determined from the pH value at the half-equivalence point using the following equation:

$$pH = pK_a = -\log K_a.$$

Describe how you would determine the acid dissociation constant using simple titration. You are provided with the following chemicals and apparatus:

- 100 cm³ of 1.00 mol dm⁻³ ethanoic acid solution;
- 100 cm³ of 1.00 mol dm⁻³ NaOH solution;
- pH meter; and
- Standard glassware for titration.

Q Where should we start thinking?

A: (i) What is the purpose of the plan?
— To determine the acid dissociation constant.

(ii) What do you need to know in order to determine the acid dissociation constant?
— We need to find out the pH at the half-equivalence point.

(iii) How do you find out the pH at the half-equivalence point?
— We need to have the titration curve. This would mean that we need to monitor the pH value using the pH meter during the whole titration course.

(iv) What would you do with the pH values that are obtained?
— We need to plot a graph of pH versus volume of NaOH solution that is added and then obtain the equivalence point. From there, we would be able to get the pH at the half-equivalence point.

(v) How would you know that you have reached the equivalence point?
— For this experiment, it is simple. Since the concentrations of both the acid and NaOH are the same, if we pipette 25 cm^3 of the acid solution, we would need 25 cm^3 of the base for complete neutralization.

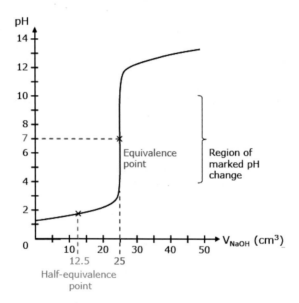

Procedure for titration:

(i) Set up the 50 cm^3 *burette* containing 1.00 mol dm^{-3} NaOH solution.
(ii) *Pipette 25.0 cm^3* of the ethanoic acid solution into a *conical flask*.
(iii) Insert a *pH probe* into the solution and take the *initial pH reading*.
(iv) Take the *initial burette reading*. *Titrate* the sample solution *against the NaOH solution*. *Swirl* continuously during the addition of the titrant.
(v) *Take the pH value for every 1 cm^3* of NaOH solution added *until 30 m^3 of NaOH solution has been added*.
(vi) *Plot a graph* of pH versus volume of NaOH solution added.
(vii) *Repeat the experiment* to check for *reliability of results*.

Q If we titrated the sodium hydroxide against the weak acid instead, would we still be able to determine the pK_a value at the half-equivalence point?

A: If you titrate a weak acid against a strong base, the pK_a value can be found at the half-equivalence point because at the half-equivalence point, the concentration of the conjugate acid (CH_3COOH) is the same as the concentration of conjugate base (CH_3COO^-). Hence according to the Henderson–Hasselbach equation:

$$pH = pK_a + \log \frac{[\text{conjugate base}]}{[\text{conjugate acid}]} = pK_a + \log 1 = pK_a.$$

But if you titrate the strong base against the weak acid instead, then at the half-equivalence point, half the amount of the strong base has reacted, generating an equal amount of the conjugate base. But there is no extra conjugate acid present because you have not continued adding the weak acid from the burette. So, there is no conjugate acid present, the ratio of $\frac{[\text{conjugate base}]}{[\text{conjugate acid}]} \neq 1$. At $\frac{[\text{conjugate base}]}{[\text{conjugate acid}]} = 1$, it is the point at twice the equivalence point:

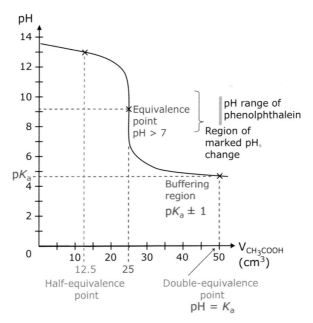

34 *Understanding Experimental Planning for Advanced Level Chemistry*

Note: In addition, such a planning process can also be used for the following similar cases:

(i) Determine the base dissociation constant, K_b, of the weak base NH_3/CH_3COO^-:

$$NH_3(aq) + H_2O(l) \rightleftharpoons NH_4^+(aq) + OH^-(aq), \quad K_b = \frac{[NH_4^+][OH^-]}{[NH_3]} \quad \text{or}$$

$$CH_3COO^-(aq) + H_2O(l) \rightleftharpoons CH_3COOH(aq) + OH^-(aq),$$

$$K_b = \frac{[CH_3COOH][OH^-]}{[CH_3COO^-]}$$

(Note: If the weak base, NH_3/CH_3COO^-, is titrated against the strong acid, then the pK_a value of NH_4^+/CH_3COOH is determined from the pH value at the half-equivalence point. To calculate the pK_b of NH_3/CH_3COO^-, simply use $pK_a + pK_b = 14$.

But if the strong acid is titrated against the weak base, NH_3/CH_3COO^-, then the pK_a value of NH_4^+/CH_3COOH is determined from the pH value at the double-equivalence point.)

(ii) Determine the acid dissociation constant, K_a, of the weak acid NH_4^+:

$$NH_4^+(aq) + H_2O(l) \rightleftharpoons NH_3(aq) + H_3O^+(aq), \quad K_a = \frac{[NH_3][H_3O^+]}{[NH_4^+]}.$$

(Note: If the weak acid is titrated against the strong base, then the pK_a value of NH_4^+ is determined from the pH value at the half-equivalence point.

But if the strong base is titrated against the weak acid, then the pK_a value of NH_4^+ is determined from the pH value at the double-equivalence point.)

1.9 Determine the Value of n in $(COOH)_2 \cdot nH_2O$ Using Acid–Base Titration

The Task:

You are given a weak dibasic acid known as ethanedioic acid, $(COOH)_2 \cdot nH_2O$. Describe how you would determine the value of n using simple titration. You are provided with the following chemicals and apparatus:

- 5 g of $(COOH)_2 \cdot nH_2O$;
- 0.100 mol dm^{-3} NaOH solution;
- 250 cm^3 volumetric flask/graduated flask; and
- Standard glassware for titration.

Q Where should we start thinking?

A: (i) What is the purpose of the plan?
— To determine the value of n in $(COOH)_2 \cdot nH_2O$.

(ii) What do you need to know in order to determine the value of n in $(COOH)_2 \cdot nH_2O$?
— We need to know the number of moles of $(COOH)_2 \cdot nH_2O$ in a fixed mass of the sample that we have weighed as the amount in moles of $(COOH)_2 \cdot nH_2O$ is calculated as follows:

Amount of $(COOH)_2 \cdot nH_2O$ in moles

$$= \frac{\text{Mass of } (COOH)_2 \cdot nH_2O \text{ used}}{\text{Relative molecular mass of } (COOH)_2 \cdot nH_2O}$$

$$= \frac{\text{Mass of } (COOH)_2 \cdot nH_2O \text{ used}}{90 + 18n}.$$

Hence, if both the amount of $(COOH)_2 \cdot nH_2O$ in moles and the mass of $(COOH)_2 \cdot nH_2O$ used are known, the value of n can be easily calculated.

(iii) How can you then determine the number of moles of $(COOH)_2 \cdot nH_2O$?
— We can make use of the following balanced equation for the reaction between $(COOH)_2 \cdot nH_2O$ and NaOH:

$2NaOH(aq) + (COOH)_2(aq) \rightarrow (COONa)_2(aq) + 2H_2O(l)$.

Thus, if we know the number of moles of NaOH that is used for the titration, we would be able to deduce the number of moles of $(COOH)_2$ that is present in the sample mass. Hence, the value of n in $(COOH)_2 \cdot nH_2O$ can be obtained as one mole of $(COOH)_2 \cdot nH_2O$ contains one mole of $(COOH)_2$ and n moles of H_2O.

(iv) But can you titrate the solid $(COOH)_2 \cdot nH_2O$ directly?
— No, we need to dissolve the $(COOH)_2 \cdot nH_2O$ in a known amount of water first.

(v) But how would you know how much of the sample to weigh and dissolve in how much of water to form the solution?

— Well, if we know the number of moles of NaOH used, we would be able to find out the number of moles of $(COOH)_2$ present in the 25 cm^3 of the solution being pipetted. Knowing this, we would be able to determine the mass of $(COOH)_2$ in the 25 cm^3 of the solution pipetted. Hence, we would be able to find out the mass of the sample that we need to measure in order to make 250 cm^3 of the solution.

(vi) Now, what is the expected volume of the NaOH solution to be used during titration?

— We can assume that we will use about 25 cm^3 of the NaOH solution to react with 25 cm^3 of the $(COOH)_2$ solution.

(vii) Why do you need to make 250 cm^3 of the sample solution?

— This is to allow us to repeat the titration till we get consistent readings of ± 0.10 cm^3.

Pre-Experimental Calculations:

$$2NaOH(aq) + (COOH)_2(aq) \rightarrow (COONa)_2(aq) + 2H_2O(l)$$

Assuming that 25 cm^3 of 0.100 mol dm^{-3} NaOH solution is used:

Amount of NaOH used = $\frac{25}{1000} \times 0.100 = 2.5 \times 10^{-3}$ mol

Assuming that 25 cm^3 of $(COOH)_2$ is pipetted:

Amount of $(COOH)_2$ present = $\frac{1}{2} \times 2.5 \times 10^{-3} = 1.25 \times 10^{-3}$ mol

Molar mass of $(COOH)_2 = 2(12.0 + 32.0 + 1.0) = 90.0$ g mol^{-1}

Mass of $(COOH)_2$ present in 25 cm$^3 = 1.25 \times 10^{-3} \times 90.0 = 0.1125$ g

Assuming that the value of n is zero, then the mass of $(COOH)_2$ is the "same" as the mass of $(COOH)_2 \cdot nH_2O$.

Hence, mass of the sample to be dissolved in 250 cm^3 water in the volumetric flask = $0.1125 \times 10 = 1.125$ g.

Why are we assuming that the value of *n* to be zero for $(COOH)_2 \cdot nH_2O$ in the above calculation?

A: By assuming that the value of *n* is zero, we would never exceed the assumed 25 cm³ of 0.100 mol dm⁻³ NaOH solution needed from the burette.

The Procedure:

Procedure for making a solution containing $(COOH)_2$:

(i) *Weigh* accurately *1.12 g* of the sample.
(ii) *Dissolve* the sample in about *50 cm³ of water in a beaker*.
(iii) *Transfer* the solution into the *volumetric flask* after ensuring that all the solids have dissolved. Rinse the beaker with water *a few times* and *transfer the washings* into the volumetric flask, to ensure quantitative transfer.
(iv) *Top up* the solution in the volumetric flask to the *graduation mark using a dropper*. *Shake* the flask well to get a *homogeneous solution*.

Procedure for titration:

(i) Set up the *50 cm³ burette* containing 0.100 mol dm⁻³ NaOH solution.
(ii) *Pipette 25.0 cm³* of the sample solution into a *conical flask*.
(iii) Add two drops of *phenolphthalein indicator* into the conical flask.
(iv) Take the *initial burette reading*. *Titrate* the sample solution *against the NaOH solution*. *Swirl* continuously during the addition of the titrant.
(v) Toward the *end-point*, add the NaOH solution *dropwise* and *swirl*. Stop the addition when *one drop* of the titrant causes the indicator to change from *colorless to pink*.
(vi) Take the *final burette reading* and calculate the *titer volume*.
(vii) *Repeat titrations* until the titer volumes are within *±0.10 cm³ consistency* (i.e., get at least two titer volumes that are within 0.10 cm³ of each other).
(viii) *Repeat the experiment* to check for *reliability of results*.

> **Q** Can we use methyl orange as an indicator instead of phenolphthalein?

A: This is a weak acid–strong base titration. As the end-point of the titration is above pH 7 due to basic hydrolysis of the conjugate base, phenolphthalein is a more suitable indicator as its working range lies within pH 8.3–10. For more details, please refer to *Understanding Advanced Physical Inorganic Chemistry* by J. Tan and K. S. Chan.

Note: In addition, such a planning process can also be used for the following similar cases:
(i) Determine the value of n for an organic acid, H_nA, with a relative molecular mass of 46.
(ii) Determine the value of n for a soluble base, $M(OH)_n$, with a relative molecular mass of 56.

> **Q** Can you cite an example to show how the value of n for an organic acid, H_nA, with a relative molecular mass of 46 can be determined?

A: Let's say: 0.046 g of H_nA requires 25.00 cm^3 of 0.040 mol dm^{-3} NaOH solution for a complete reaction. Determine the value of n for an organic acid, H_nA.

$$\text{Amount of } H_nA \text{ used} = \frac{0.046}{46} = 0.010 \text{ mol}$$
$$\text{Amount of NaOH used} = \frac{25}{1000} \times 0.4 = 0.01 \text{ mol}$$

Since the mole ratio of H_nA : NaOH = 1:1, $n = 1$.

1.10 Determine the Value of n in $(COOH)_2 \cdot nH_2O$ Using Redox Titration

> *The Task:*
>
> Ethanedioic acid, $(COOH)_2 \cdot nH_2O$, undergoes a redox reaction with acidified $KMnO_4$:
>
> $$2MnO_4^- + 5(COOH)_2 + 6H^+ \rightarrow 2Mn^{2+} + 10CO_2 + 8H_2O.$$
>
> Describe how you would determine the value of n using a simple redox titration. You are provided with the following chemicals and apparatus:

- 12 g of $(COOH)_2 \cdot nH_2O$;
- 0.010 mol dm^{-3} $KMnO_4$ solution;
- 0.05 mol dm^{-3} H_2SO_4 solution;
- 250 cm^3 volumetric flask/graduated flask; and
- Standard glassware for titration.

Q Where should we start thinking?

A: (i) What is the purpose of the plan?
— To determine the value of n in $(COOH)_2 \cdot nH_2O$.

(ii) What do you need to know in order to determine the value of n in $(COOH)_2 \cdot nH_2O$?
— We need to know the number of moles of $(COOH)_2 \cdot nH_2O$ in a fixed mass of the sample that we have weighed as the amount in moles of $(COOH)_2 \cdot nH_2O$ is calculated as follows:

Amount of $(COOH)_2 \cdot nH_2O$ in moles

$$= \frac{\text{Mass of } (COOH)_2 \cdot nH_2O \text{ used}}{\text{Relative molecular mass of } (COOH)_2 \cdot nH_2O}$$

$$= \frac{\text{Mass of } (COOH)_2 \cdot nH_2O \text{ used}}{90 + 18n}.$$

Hence, if both the amount of $(COOH)_2 \cdot nH_2O$ in moles and the mass of $(COOH)_2 \cdot nH_2O$ used are known, the value of n can be easily calculated.

(iii) How do you then determine the number of moles of $(COOH)_2 \cdot nH_2O$?
— We can make use of the following balanced equation between the reaction of $(COOH)_2 \cdot nH_2O$ and MnO_4^-:

$$2MnO_4^- + 5(COOH)_2 + 6H^+ \rightarrow 2Mn^{2+} + 10CO_2 + 8H_2O.$$

Thus, if we know the number of moles of MnO_4^- used for the titration, we would be able to deduce the number of moles of $(COOH)_2$ present in the sample mass. Hence, the value of n in $(COOH)_2 \cdot nH_2O$ can be obtained as one mole of $(COOH)_2 \cdot nH_2O$ contains one mole of $(COOH)_2$ and n moles of H_2O.

(iv) But can you titrate the solid $(COOH)_2 \cdot nH_2O$ directly?
— No, we need to dissolve the $(COOH)_2 \cdot nH_2O$ in a known amount of water first.

(v) But how would you know how much of the sample to weigh and dissolve in how much of water to form the solution?
— Well, if we know the number of moles of MnO_4^- that is used, we would be able to find out the number of moles of $(COOH)_2$ that is present in the 25 cm³ of solution that is being pipetted. Knowing this, we would be able to determine the mass of $(COOH)_2$ that is present in the 25 cm³ of solution pipetted. Hence, we would be able to find out the mass of the sample that we need to measure in order to make 250 cm³ of the solution.

(vi) Now, what is the expected volume of the MnO_4^- solution that you would use during titration?
— We can assume that about 25 cm³ of the MnO_4^- solution will react with 25 cm³ of the $(COOH)_2$ solution.

(vii) Why do you need to make 250 cm³ of sample solution?
— This is to allow us to repeat the titration till we get consistent readings of ± 0.10 cm³.

Pre-Experimental Calculations:

$$2MnO_4^- + 5(COOH)_2 + 6H^+ \rightarrow 2Mn^{2+} + 10CO_2 + 8H_2O$$

Assuming that 25 cm³ of 0.100 mol dm⁻³ MnO_4^- solution is used:

Amount of MnO_4^- used = $\frac{25}{1000} \times 0.010 = 2.5 \times 10^{-4}$ mol

Amount of H^+ needed = $\frac{6}{2} \times 2.5 \times 10^{-4} = 7.5 \times 10^{-4}$ mol

Amount of H_2SO_4 needed = $\frac{1}{2} \times 7.5 \times 10^{-4} = 3.75 \times 10^{-4}$ mol

Volume of H_2SO_4 needed = $\frac{3.75 \times 10^{-4}}{0.05} = 7.5 \times 10^{-3}$ dm³ = 7.5 cm³

Assuming that 25 cm³ of $(COOH)_2$ is pipetted:

Amount of $(COOH)_2$ present = $\frac{5}{2} \times 2.5 \times 10^{-4} = 6.25 \times 10^{-4}$ mol

Molar mass of $(COOH)_2 = 2(12.0 + 32.0 + 1.0) = 90.0$ g mol⁻¹

Mass of $(COOH)_2$ present in 25 cm³ = $6.25 \times 10^{-4} \times 90.0 = 0.5625$ g

Assuming that the value of n is zero, then the mass of $(COOH)_2$ is the "same" as the mass of $(COOH)_2 \cdot nH_2O$.

Hence, mass of the sample to be dissolved in 250 cm³ water in the volumetric flask = $0.5625 \times 10 = 5.625$ g.

Q Did you notice the similarities between Sections 1.9 and 1.10?

A: Yes...

The Procedure:

Procedure for making a solution containing $(COOH)_2$:

  (i) *Weigh accurately 5.62 g of the sample.*
  (ii) *Dissolve the sample in about 50 cm³ of water in a beaker.*
  (iii) *Transfer the solution into the volumetric flask after ensuring that all the solids have dissolved. Rinse the beaker with water a few times and transfer the washings into the volumetric flask, to ensure quantitative transfer.*
  (iv) *Top up the solution in the volumetric flask to the graduation mark using a dropper. Shake the flask well to get a homogeneous solution.*

Procedure for titration:

  (i) Set up the 50 cm³ *burette* containing 0.010 mol dm⁻³ *acidified* $KMnO_4$ solution.
  (ii) *Pipette* 25.0 cm³ of the sample solution into a *conical flask*.
  (iii) Use a *measuring cylinder*, measure 8 cm³ of 0.05 mol dm⁻³ H_2SO_4 solution into the conical flask.
  (iv) Take the *initial burette reading*. *Titrate* the sample solution *against the* $KMnO_4$ *solution*. *Swirl* continuously during the addition of the titrant.
  (v) Toward the *end-point*, add the $KMnO_4$ solution *dropwise* and *swirl*. Stop the addition when *one drop* of titrant causes the solution to change from *colorless to pink*.
  (vi) Take the *final burette reading* and calculate the *titer volume*.

(vii) *Repeat titrations* until the titer volumes are within ± 0.10 cm^3 *consistency* (i.e., get at least two titer volumes that are within 0.10 cm^3 of each other).

(viii) *Repeat the experiment* to check for *reliability of results*.

> **Q** Why do we need to add H_2SO_4 solution before the titration?

A: Some redox reactions, such as the one above, need an acidic medium. Hence, additional H^+ ions need to be added.

> **Q** Why do you use a measuring cylinder to measure the H_2SO_4 solution? Shouldn't a more accurate apparatus, such as a burette, be used?

A: Since the H_2SO_4 solution that is added is already in excess, it is alright to use a measuring cylinder to measure the volume as accuracy is no longer an issue.

> **Q** Can we use HCl or HNO_3 as the acid instead?

A: No! HCl can undergo a redox reaction with $KMnO_4$ while HNO_3 itself is an oxidizing agent. As for H_2SO_4, the SO_4^{2-} ion usually does not participate in the redox reaction, hence it is a better acidification tool.

> **Q** What causes the pink coloration at the end-point?

A: The pink coloration arises because of one excess drop of acidified $KMnO_4$ solution. Thus, for a redox titration involving acidified $KMnO_4$, an external indicator is not needed. And usually, the acidified $KMnO_4$ solution "sits" in the burette as we need that excess drop of purple acidified $KMnO_4$ to "signal" the end-point of titration.

Note: In addition, such a planning process can also be used for the following similar cases:

(i) Determine the value of n for $Fe(NH_4)_2(SO_4)_2 \cdot nH_2O$ through redox titration using acidified $KMnO_4$:

$$MnO_4^-(aq) + 8H^+(aq) + 5Fe^{2+}(aq) \rightarrow Mn^{2+}(aq) + 5Fe^{3+}(aq) + 4H_2O(l).$$

(ii) Determine the concentration of a H_2O_2 solution through redox titration using acidified $KMnO_4$:

$$2MnO_4^-(aq) + 6H^+(aq) + 5H_2O_2(aq) \rightarrow 2Mn^{2+}(aq) + 5O_2(g) + 8H_2O(l).$$

(iii) Determine the percentage of Fe^{2+} in an iron pill through redox titration using acidified $K_2Cr_2O_7$:

$$Cr_2O_7^{2-}(aq) + 14H^+(aq) + 6Fe^{2+}(aq) \rightarrow 2Cr^{3+}(aq) + 6Fe^{3+}(aq) + 7H_2O(l).$$

1.11 Determine the Value of n in $CuSO_4 \cdot nH_2O$ Using Redox Titration

The Task:

Copper(II) ions react with iodide ions to form a cream precipitate of copper(I) iodide in a brown solution of iodine:

$$2Cu^{2+}(aq) + 4I^-(aq) \rightarrow 2CuI(s) + I_2(aq).$$

Iodine reacts with thiosulfate ions according to the equation below:

$$I_2(aq) + 2S_2O_3^{2-}(aq) \rightarrow 2I^-(aq) + S_4O_6^{2-}(aq).$$

Describe how you would determine the value of n in $CuSO_4 \cdot nH_2O$, using a simple redox titration. You are provided with the following chemicals and apparatus:

- 8 g of $CuSO_4 \cdot nH_2O$;
- 1.0 mol dm^{-3} potassium iodide, KI, solution;
- 0.100 mol dm^{-3} sodium thiosulfate, $Na_2S_2O_3$, solution;
- 250 cm^3 volumetric flask/graduated flask; and
- Standard glassware for titration.

Q Where should we start thinking?

A: (i) What is the purpose of the plan?
— To determine the value of n in $CuSO_4 \cdot nH_2O$.

(ii) What do you need to know in order to determine the value of n in $CuSO_4 \cdot nH_2O$?

— We need to know the number of moles of $CuSO_4 \cdot nH_2O$ in a fixed mass of the sample that we have weighed as the amount in moles of $CuSO_4 \cdot nH_2O$ is calculated as follows:

Amount of $CuSO_4 \cdot nH_2O$ in moles

$$= \frac{\text{Mass of } CuSO_4 \cdot nH_2O \text{ used}}{\text{Relative molecular mass of } CuSO_4 \cdot nH_2O}$$

$$= \frac{\text{Mass of } CuSO_4 \cdot nH_2O \text{ used}}{157.6 + 18n}.$$

Hence, if the both the amount of $CuSO_4 \cdot nH_2O$ in moles and the mass of $CuSO_4 \cdot nH_2O$ used are known, the value of n can be easily calculated.

(iii) How can you then determine the number of moles of $CuSO_4 \cdot nH_2O$?

— As the Cu^{2+} ions react with I^- to give us a specific amount of I_2:

$$2Cu^{2+}(aq) + 4I^-(aq) \rightarrow 2CuI(s) + I_2(aq),$$

we can determine the amount of I_2 by using the following balanced equation for the reaction between of I_2 and $S_2O_3^{2-}$:

$$I_2(aq) + 2S_2O_3^{2-}(aq) \rightarrow 2I^-(aq) + S_4O_6^{2-}(aq).$$

Thus, if we know the number of moles of $S_2O_3^{2-}$ used for the titration, we would be able to deduce the number of moles of I_2 formed and thus the amount of Cu^{2+} present in the sample. Therefore, the value of n in $CuSO_4 \cdot nH_2O$ can be obtained as one mole of $CuSO_4 \cdot nH_2O$ contains one mole of $CuSO_4$ and n moles of H_2O.

(iv) But can you titrate the solid $CuSO_4 \cdot nH_2O$ directly?

— No, we need to dissolve the $CuSO_4 \cdot nH_2O$ in a known amount of water first.

(v) But how would you know how much of the sample to weigh and dissolve in how much of water to form the solution?

— Well, if we know the number of moles of $S_2O_3^{2-}$ that has been used, we would be able to find out the number of moles of Cu^{2+} in the 25 cm^3 of solution that has been pipetted. Knowing this, we would be able to determine the mass of $CuSO_4 \cdot nH_2O$ in the 25 cm^3 of solution that has been pipetted. Hence, we would be able to find out the mass of the sample that we need to measure in order to have 250 cm^3 of the solution.

(vi) Now, what is the expected volume of the $S_2O_3^{2-}$ solution that is used during titration?
— We can assume that about $25\,cm^3$ of the $S_2O_3^{2-}$ solution will react with $25\,cm^3$ of the Cu^{2+} solution.

(vii) Why do you need to make $250\,cm^3$ of sample solution?
— This is to allow us to repeat the titration till we get consistent readings of $\pm 0.10\,cm^3$.

Pre-Experimental Calculations:

$$I_2(aq) + 2S_2O_3^{2-}(aq) \rightarrow 2I^-(aq) + S_4O_6^{2-}(aq)$$

Assuming that $25\,cm^3$ of $0.100\,mol\,dm^{-3}$ $S_2O_3^{2-}$ solution is used:

Amount of $S_2O_3^{2-}$ used $= \frac{25}{1000} \times 0.100 = 2.5 \times 10^{-3}$ mol

Amount of I_2 present $= \frac{1}{2} \times 2.5 \times 10^{-3} = 1.25 \times 10^{-3}$ mol

$$2Cu^{2+}(aq) + 4I^-(aq) \rightarrow 2CuI(s) + I_2(aq)$$

Amount of Cu^{2+} present $= 2 \times 1.25 \times 10^{-3} = 2.5 \times 10^{-3}$ mol

Amount of I^- needed $= 4 \times 1.25 \times 10^{-3} = 5.0 \times 10^{-3}$ mol

Volume of $1.0\,mol\,dm^{-3}$ KI solution needed $= \frac{5.0 \times 10^{-3}}{1.0} = 5.0 \times 10^{-3}\,dm^3 = 5\,cm^3$

Assuming that $25\,cm^3$ of $CuSO_4$ solution is pipetted:

Amount of $CuSO_4$ present $= 2.5 \times 10^{-3}$ mol

Molar mass of $CuSO_4 = 63.5 + 32.1 + 4(16.0) = 157.6\,g\,mol^{-1}$

Mass of $CuSO_4$ present in $25\,cm^3 = 2.5 \times 10^{-3} \times 157.6 = 0.394\,g$

Assuming that the value of n is zero, then the mass of $CuSO_4$ is the "same" as the mass of $CuSO_4 \cdot nH_2O$.

Hence, mass of the sample to be dissolved in $250\,cm^3$ water in the volumetric flask $= 0.394 \times 10 = 3.94\,g$.

Q Did you notice the similarities between Sections 1.9, 1.10 and 1.11?

A: Yes. The fundamental idea of such planning is to obtain the number of moles of an "unknown" through titration with the help of a balanced equation.

The Procedure:

Procedure for making a solution containing $CuSO_4$:
   (i) *Weigh* accurately *3.94 g of the solid* $CuSO_4 \cdot nH_2O$.
   (ii) *Dissolve* the sample in about *50 cm^3 of water in a beaker*.
   (iii) *Transfer* the solution into the *volumetric flask* after ensuring that *all the solids have dissolved. Rinse the beaker with water a few times* and *transfer the washings* into the volumetric flask, to ensure *quantitative transfer*.
   (iv) *Top up* the solution in the volumetric flask to the *graduation mark* using a *dropper. Shake* the flask well to get a *homogeneous solution*.

Procedure for titration:

   (i) Set up the 50 cm^3 *burette* containing 0.100 mol dm^{-3} *sodium thiosulfate solution*.
   (ii) *Pipette 25.0 cm^3* of the sample solution into a *conical flask*.
   (iii) Use a *measuring cylinder*, add *6 cm^3* of 1.0 mol dm^{-3} KI solution into the conical flask.
   (iv) Take the *initial burette reading. Titrate* the sample solution *against the sodium thiosulfate solution. Swirl* continuously during the addition of the titrant.
   (v) When the solution in the conical flask turns *pale yellow*, add three drops of *starch solution* and then continue the titration.
   (vi) Stop the addition of the titrant when *one drop of titrant* causes the *blue-black coloration to decolorize*.
   (vii) Take the *final burette reading* and calculate the *titer volume*.
   (viii) *Repeat titrations* until the titer volumes are within ±*0.10 cm^3 consistency* (i.e., get at least two titer volumes that are within 0.10 cm^3 of each other).
   (ix) *Repeat the experiment* to check for *reliability of results*.

> **Q** Why do you use a measuring cylinder to measure the volume of the KI solution? Shouldn't a more accurate apparatus, such as a burette, be used?

A: Since the KI that is being added is already in excess, it is alright to use a measuring cylinder to measure the volume as accuracy is no longer an issue.

Q What is the purpose of adding starch solution towards the end-point of titration?

A: Iodine in the presence of excess iodide ions would form a brown I_3^- complex, which is more soluble in water:

$$I_2(aq) + I^-(aq) \rightleftharpoons I_3^-(aq) \text{ (brown coloured complex)}.$$

As thiosulfate ions are being added, iodine would be reduced, resulting in a decrease of the concentration of the iodine. As the concentration of the I_3^- complex diminishes, the resultant solution approaches a pale yellow coloration. In order to have a more accurate detection of the end-point color change, a few drops of starch solution is normally added when the solution turns pale yellow:

$$I_2(aq) + \text{starch molecules} \rightleftharpoons I_2\text{-starch complex}$$
$$\text{(blue-black coloration)}.$$

The starch molecules, which are none other than polysaccharides, form a deep blue water-insoluble complex. The deep blue coloration is formed due to iodine molecules being trapped within the polysaccharides. The formation of the iodine–starch complex is a reversible reaction. Now, as more thiosulfate solution is added, more iodine would react with the thiosulfate ions. The iodine–starch complex would then break down to release the trapped iodine molecules. Hence, a distinct color change from deep blue to colorless would be observed. This should be the point whereby the volume of titrant used is to be noted. It is because if the titrated solution is left to stand in air, the deep blue coloration may be restored. This should be ignored, as it is due to atmospheric oxidation of the excess iodide in the reaction mixture back to iodine.

 Q Why can't we add starch solution at the very beginning of the titration process?

A: Now, since the iodine is trapped in the starch molecules, adding starch solution at the start may cause many iodine molecules to be embedded in the polysaccharides. The release of iodine from the starch molecules takes time and as such, it might decrease the accuracy of the titration result.

> **Q** How does the iodine molecules get adsorb onto the starch molecules?

A: Now iodine is a non-polar diatomic molecule whereas starch molecule is made up of glucose molecules, hence it is polar in nature. The interaction between iodine and starch molecules would be van der Waals' forces of the instantaneous dipole–induced dipole (id-id) type (please refer to *Understanding Advanced Physical Inorganic Chemistry* by J. Tan and K. S. Chan for more details).

1.12 Determine the Solubility Product Constant (K_{sp}) for Ba(IO$_3$)$_2$ Using Redox Titration

The Task:

Barium iodate(V), $Ba(IO_3)_2$, is partially soluble in water:
$$Ba(IO_3)_2(s) \rightleftharpoons Ba^{2+}(aq) + 2IO_3^-(aq), \quad K_{sp} = [Pb^{2+}][IO_3^-]^2 \text{ mol}^3 \text{ dm}^{-9}.$$
The IO_3^-(aq) will react with excess KI as follows:
$$IO_3^-(aq) + 5I^-(aq) + 6H^+(aq) \rightarrow 3I_2(aq) + 3H_2O(l).$$
Iodine reacts with thiosulfate ions according to the equation below:
$$I_2(aq) + 2S_2O_3^{2-}(aq) \rightarrow 2I^-(aq) + S_4O_6^{2-}(aq).$$
Describe how you would determine the K_{sp} value of $Ba(IO_3)_2$ using a simple redox titration. You are provided with the following chemicals and apparatus:

- 8 g of solid $Ba(IO_3)_2$;
- 1.0 mol dm^{-3} potassium iodide, KI, solution;
- 0.100 mol dm^{-3} sodium thiosulfate, Na$_2$S$_2$O$_3$, solution;
- 250 cm^3 volumetric flask/graduated flask; and
- Standard glassware for titration.

> **Q** Where should we start thinking?

A: (i) What is the purpose of the plan?
— To determine the K_{sp} value of $Ba(IO_3)_2$.

(ii) What do you need to do then?
— We need to dissolve some $Ba(IO_3)_2$ in pure water first.
(iii) What do you need to know in order to determine how much of the $Ba(IO_3)_2$ has dissolved?
— We can filter the mixture first, then pipette the filtrate and add some excess KI solution to the filtrate:
$$IO_3^-(aq) + 5I^-(aq) + 6H^+(aq) \rightarrow 3I_2(aq) + 3H_2O(l).$$
The iodine that is formed can be titrated with the given standard sodium thiosulfate solution.
(iv) How can you then determine the number of moles of $Ba(IO_3)_2$?
— We can make use of the following balanced equation for the reaction between I_2 and $S_2O_3^{2-}$:
$$I_2(aq) + 2S_2O_3^{2-}(aq) \rightarrow 2I^-(aq) + S_4O_6^{2-}(aq).$$
Thus, if we know the number of moles of $S_2O_3^{2-}$ used for the titration, we would be able to deduce the number of moles of $I_2(aq)$ formed. This in turn would give us the number of moles of $IO_3^-(aq)$ present and thus the number of moles of $Ba(IO_3)_2$ dissolved.

Q Can we do a pre-calculation to find out how much $Na_2S_2O_3$ is needed for the titration?

A: No! As you do not know the solubility of the $Ba(IO_3)_2$, you can't do a pre-calculation. Anyway, the whole exercise here is to let you be aware of the thinking process that is involved when we approach the question.

The Procedure:

Procedure for dissolving the insoluble $Ba(IO_3)_2$:
(i) Use a 50 cm³ *burette* to introduce *100 cm³* of deionized water into a *clean and dry conical flask*.
(ii) Introduce some *solid $Ba(IO_3)_2$*, and *stir* the solution until *some solid remains undissolved*.
(iii) *Filter* the mixture into a *clean and dry conical flask*.

> **Q** Why must we filter the mixture into a clean and dry conical flask?

A: If the conical flask is not dry, the water present would dilute the filtrate. This would affect the concentration of the $Ba(IO_3)_2$ that we are going to determine.

Procedure for titration:

(i) Set up the 50 cm³ *burette* containing the *standard $Na_2S_2O_3$ solution*.
(ii) *Pipette 25.0 cm³* of the *filtrate* into a conical flask.
(iii) Use a *measuring cylinder*, add in some *excess KI solution*.
(iv) Take the *initial burette reading. Titrate* the sample solution against the *sodium thiosulfate solution. Swirl* continuously during the addition of the titrant.
(v) When the solution in the conical flask turns *pale yellow*, add three drops of *starch solution* and then continue the titration.
(vi) Stop the addition of the titrant when *one drop* of titrant causes the blue-black coloration to decolorize.
(vii) Take the *final burette reading* and calculate the *titer volume*.
(viii) *Repeat titrations* until the titer volumes are within ±*0.10 cm³ consistency* (i.e., get at least two titer volumes that are within 0.10 cm³ of each other).
(ix) *Repeat the experiment* to check for *reliability of results*.

1.13 Determine the Partition Coefficient of a Weak Acid Through Titration

The Task:

Partition coefficient, K_{PC}, is an equilibrium constant which is temperature dependent. It is the ratio of the equilibrium concentrations of a weak acid, HX, in a mixture of two immiscible solvents. This equilibrium constant measures the relative solubility of the weak acid in these two immiscible solvents:

$$HX(aq) \rightleftharpoons HX(organic), \quad K_{PC} = \frac{[HX(organic)]}{[HX(aq)]}.$$

Planning Using Titration

Describe how you would determine the K_{PC} value of HX using a simple titration. You are provided with the following chemicals and apparatus:

- 100 cm^3 of HX solution;
- 100 cm^3 of hexane;
- A separatory/separating funnel for mixing and separating the two immiscible solvents;
- Standard NaOH solution; and
- Standard glassware for titration.

Q Where should we start thinking?

A: (i) What is the purpose of the plan?
— To determine the K_{PC} value of HX using a simple titration.

(ii) What do you need to do then?
— We need to mix a known volume of the HX solution with a known volume of hexane together. Then, separate these two immiscible solvents and titrate the acid in the aqueous medium using the standard sodium hydroxide solution.

(iii) How would you know how much of the HX is in the organic medium?
— As the initial number of moles of HX used is known, the titration result would give us the remaining HX in the aqueous medium. Take the difference of these two values and this would give us the number of moles of HX that has gone into the organic layer.

Q Can we do a pre-calculation to find out how much NaOH is needed for the titration?

A: No! As you do not know the solubility of the HX in each of the two mediums, you can't do a pre-calculation. Anyway, the whole exercise here is to let you be aware of how we can approach the question.

> **Q** How do we use the separatory funnel?

A: If we have two liquids, such as water and oil, that do not mix with each other and are separated into two distinct layers, we say that they are *immiscible*. A separating funnel is used to separate such liquids as shown below:

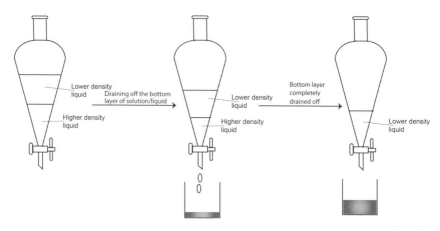

> **Q** If both of the immiscible layers are colorless, how would you know which is the aqueous layer?

A: Just take a dropper, drip one drop of water onto the top layer. If you see the water droplet moves through the top layer, then this top layer must be the organic solvent which is immiscible with water. If not, then the top layer is the aqueous layer.

The Procedure:

Procedure for partitioning HX between the two immiscible solvents:
  (i) Use a 50 cm^3 *burette* to introduce *50 cm^3* of aqueous HX into a *clean and dry separatory funnel*.
  (ii) Use another 50 cm^3 *burette* to introduce *50 cm^3* of hexane into the same separatory funnel.
  (iii) *Invert* the separatory funnel and *shake* the mixture vigorously. *Open the stop-cock* occasionally to release the pressure buildup.
  (iv) *Drain* the aqueous medium into a clean and dry conical flask.

Procedure for titration:

(i) Set up the 50 cm^3 burette containing the *standard NaOH solution*.
(ii) *Pipette 10.0 cm^3* of the *filtrate* into a conical flask.
(iii) Add two drops of *phenolphthalein indicator*.
(iv) Take the *initial burette reading*. *Titrate* the sample solution against the *sodium hydroxide solution*. *Swirl* continuously during the addition of the titrant.
(v) Stop the addition when *one drop* of the titrant causes the color of solution to turn from *colorless to pale pink*.
(vi) Take the *final burette reading* and calculate the *titer volume*.
(vii) *Repeat titrations* until the titer volumes are within ±*0.10* cm^3 consistency (i.e., get at least two titer volumes that are within 0.10 cm^3 of each other).
(viii) *Repeat the experiment* to check for reliability of results.

> **Q** How can we use the K_{PC} to check in which solvent the HX is more soluble?

A: As K_{PC} is a ratio, $K_{PC} = \dfrac{[\text{HX(organic)}]}{[\text{HX(aq)}]}$, the greater the value would mean that HX is more soluble in the organic medium, and vice versa if you have a smaller value.

1.14 Safety Precautions for Titration Experiments

You may be asked to quote some safety precautions while performing a titration experiment. Depending on the type of titration that you are performing, the following examples may be useful for you to take note:

— Always wear gloves, lab coat and safety goggles while doing experiment. For example: the ethanedioic acid handled in Section 1.9 is toxic in nature; the acids (H_2SO_4) or bases (NaOH) or even the oxidizing agents used may be corrosive in nature. So, there should be minimal direct contact with these chemicals.

- Use a pipette filler during the pipetting of solution. DO NOT use your mouth to suck up the solution.
- When filling up the burette, wear goggles and make sure you fill it up below your eye level. This to prevent spillage of the solution into your eyes.

1.15 Minimizing Experimental Errors or Increasing Reliability

You may be asked to quote some ways to improve experimental reliability while performing the titration experiment. Depending on the type of titration that you are performing, the following examples may be useful for you to take note:

- Swirling during titration is important as it ensures that the reactants are well mixed.
- Titration should be repeated until at least two consistent results are obtained, i.e., get at least two titer volumes that are within ±0.10 cm^3 of each other.
- The experiment must always be repeated to check for reliability of results. An average value should be calculated if necessary.
- Standard sodium hydroxide solution must be freshly prepared as it can absorb CO_2 from the air which affects its concentration. This would in turn affect the reliability of the titration result.

CHAPTER 2

PLANNING USING GRAVIMETRIC ANALYSIS

Gravimetric analysis refers to analytical methods that are used to determine the quantity of an analyte through the measurement of mass. The mass of the substance that is measured can be either the residue that is formed after a particular decomposition reaction or an insoluble precipitate that is formed during the isolation of an ion from the aqueous medium. A typical gravimetric analysis experiment requires the following apparatus:

— An electronic weighing balance;
— Crucible or boiling tube for decomposition reaction to take place;
— A dessicator for cooling purposes;
— Filtration system, includes, filter funnel and filter paper; and
— Drying oven or infrared lamp for drying residue.

The gravimetric analysis method is easy to carry out and needs only simple apparatus. The disadvantage is probably due to the time-consuming nature because of the heating–cooling–weighing cycle.

The experimental set-ups for gravimetric analysis can be as follows:

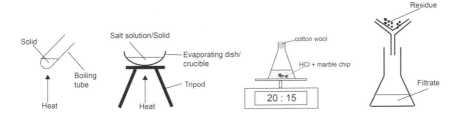

2.1 Mixture of Potassium Hydrogen Carbonate and Potassium Chloride

The Task:

A sample of potassium hydrogen carbonate, $KHCO_3$, contains 4–8% of potassium chloride, KCl. Under strong heating, the $KHCO_3$ undergoes the following decomposition:

$$2KHCO_3(s) \rightarrow K_2CO_3(s) + CO_2(g) + H_2O(g).$$

You are supposed to plan an experiment to determine the actual percentage of the potassium hydrogen carbonate in the sample with the following apparatus:

- 10 g of $KHCO_3$ and KCl mixture;
- Crucible; and
- A weighing balance.

Q Where should we start thinking?

A: (i) What is the purpose of the plan?
— To determine the percentage by mass of the potassium hydrogen carbonate in the mixture.

(ii) What do you need to know in order to determine the mass of potassium hydrogen carbonate?
— We need to know the number of moles of potassium hydrogen carbonate present in a fixed mass of the sample that we have measured.

(iii) How can you then determine the number of moles of potassium hydrogen carbonate?
— We can make use of the following decomposition equation:
$$2KHCO_3(s) \rightarrow K_2CO_3(s) + CO_2(g) + H_2O(g).$$
Thus, if we know the mass loss due to the CO_2 and H_2O, we would be able to determine the number of moles of CO_2 and H_2O. And hence, we can deduce the number of moles of $KHCO_3$ present in the sample that we have weighed.

(iv) Can we use up all the sample that is given to us at one go?
— It is very important to repeat your experiment for better reliability of results, so you should not use up all the sample at one go.

Unlike other Group 2 hydrogen carbonates, which will decompose to give the oxide, why do Group 1 hydrogen carbonates decompose to give the carbonate instead?

A: Group 2 hydrogen carbonates decomposed to give the oxide is because of the high charge density of the Group 2 cations. The charge densities of the Group 1 cations are not high enough to polarize the intramolecular covalent bonds within the HCO_3^- ion, hence you don't get the oxide formed. For details, refer to *Understanding Advanced Physical Inorganic Chemistry* by J. Tan and K. S. Chan.

The Procedure:

The general procedure for gravimetric analysis experiments involves taking the mass of the sample before and after heating or taking the mass of a dried residue after filtration. The accuracy in the measurement of the mass of a substance in a gravimetric experiment is important. Thus, the procedure of a typical gravimetric experiment consists of a series of steps which inform the student: (1) which step should come first; (2) what apparatus should he/she use; (3) what is the quantity of substance that he/she should measure; and (4) if needed, the reaction conditions such as temperature and pressure.

Procedure for the heating:

(i) *Weigh* accurately the mass of a *dry crucible/boiling tube*.
(ii) *Weigh* accurately 5.0 g of the sample. *Record the total mass.*
(iii) *Heat* the sample *gently and then strongly* for 10 min. Ensure that *all the solid is exposed to the heat.*
(iv) *Cool* the crucible/boiling tube in a *dessicator*. Then, *weigh the crucible and its content.*
(v) *Repeat* the process of heating, cooling, and weighing till *constant mass* is achieved.
(vi) *Repeat the experiment* for better *reliability of results.*

> **Q** Is it always better to use a crucible?

A: It all depends. For example, if you heat something that gives off water vapor, then an open crucible is preferred as the water vapor will condense on the cooler part of the boiling tube. The condensed water may drip back and cause the boiling tube to crack. But if you are decomposing something that produces too much gases, then boiling tube is better because an open crucible is more prone to material loss caused by the evolution of gases.

> **Q** Can we use a sample of smaller mass for the experiment?

A: Remember from Chapter 1, we learned that smaller measurement would incur a greater amount of error in the measurement? Since we are given 10 g of the sample, we can simply use 5 g each for our experiment. In addition, if a sample of smaller mass is used, the change in mass after heating is small, and there will be a greater percentage error in measuring the change in mass.

> **Q** So, we should use a sample of greater mass instead?

A: This may not be good either because if a sample of greater mass is used, the decomposition may not be completed as it is difficult to expose all of the sample to the heat.

> **Q** Why do we need to heat the sample gently first? Why can't we immediately heat the sample strongly?

A: Well, if you immediately heat the sample strongly, the sudden decomposition would cause many gases to evolve. This would cause the powdered sample to spatter and escape from the crucible.

> **Q** Why do we need to cool down the hot object before weighing?

A: The heat emitted would cause a convection or draft to form in the balance; this would cause the reading taken to be inaccurate.

> **Q** Why must the cooling be done in a dessicator?

A: The residue may absorb moisture during the cooling stage. We would get inaccurate results if this happens.

> **Q** Why do you need to expose all the solid to the heat?

A: Uneven heating would result in incomplete decomposition.

> **Q** What is the meaning of 'heat to constant mass?'

A: The progress of thermal decomposition is monitored by the loss in mass of the sample. Thermal decomposition is deemed to have been completed if you get a constant mass of the residue after the weighing, heating and cooling cycle.

Result and Analysis:

Mass of empty crucible/g	X
Mass of crucible and sample/g	Y
Mass of crucible and its contents	
After first heating/g	W
After second heating/g	Z
After third heating/g	Z

$$2KHCO_3(s) \rightarrow K_2CO_3(s) + CO_2(g) + H_2O(g)$$

Mass of $KHCO_3$ + KCl $\quad = (Y - X)$ g
Mass of residue (K_2CO_3 + KCl) $= (Z - X)$ g
Mass of CO_2 + H_2O loss $\quad = (Y - X) - (Z - X) = (Y - Z) = A$ g

Since for every one mole of CO_2 lost, there is also one mole H_2O lost, hence the sum of the molar mass of $CO_2 + H_2O = 44 + 18 = 62$ g mol^{-1}
Amount of $CO_2 + H_2O$ loss $= \frac{A}{62} = B$ mol
Amount of $KHCO_3 = 2 \times$ Amount of $CO_2 + H_2O$ loss $= 2B$ mol
Mass of $KHCO_3 = 2B \times$ Molar mass of $KHCO_3 = C$ g
Percentage of $KHCO_3$ in the sample $= \frac{C}{(Y-X)} \times 100\%$.

> **Q** In what other decomposition reactions can the above planning outline be applied to?

A: Many! For example:
- Determine the value of n of $CuSO_4 \cdot nH_2O$:
$$CuSO_4 \cdot nH_2O(s) \rightarrow CuSO_4(s) + nH_2O(g).$$
- Determine the relative molecular mass of **X** in **X**.$5H_2O$:
$$X.5H_2O(s) \rightarrow X(s) + 5H_2O(g).$$
- Determine the identity of the metal cation in MCO_3:
$$MCO_3(s) \rightarrow MO(s) + CO_2(g).$$
- Determine the purity of $Fe(NO)_3$ in a sample:
$$4Fe(NO_3)_3(s) \rightarrow 2Fe_2O_3(s) + 12NO_2(g) + 3O_2(g).$$
- Determine the percentage composition of $MgCO_3$ contaminated with Na_2CO_3:
$$MgCO_3(s) \rightarrow MgO(s) + CO_2(g);$$
$$Na_2CO_3(s) \rightarrow \text{No decomposition.}$$

Did you notice that some of the above experiments have been discussed using titration planning in Chapter 1?

> **Q** Can you roughly show us how to calculate the value of n of $CuSO_4 \cdot nH_2O$?

A: $$CuSO_4 \cdot nH_2O(s) \rightarrow CuSO_4(s) + nH_2O(g)$$
Mass of the residue gives the number of moles of $CuSO_4 = a$ mol
Loss in mass gives the number of moles of $H_2O = b$ mol
Moles ratio of $CuSO_4 : H_2O = 1 : n = a : b$

Planning Using Gravimetric Analysis 61

$$\Rightarrow \frac{1}{n} = \frac{a}{b}$$

$$\Rightarrow n = \frac{b}{a}.$$

OR

Mass of empty crucible/g	X
Mass of crucible and sample/g	Y
Mass of crucible and its contents	
After first heating/g	W
After second heating/g	Z
After third heating/g	Z

Mass of residue ($CuSO_4$) = (Z − X) g
Mass of $CuSO_4 \cdot nH_2O$ = (Y − X) g
Molar mass of $CuSO_4$ = 63.5 + 32.1 + 4(12.0) = 143.6 g mol^{-1}
Amount of $CuSO_4 = \frac{(Z-X)}{143.6} = D$ mol
Amount of $CuSO_4 \cdot nH_2O$ = Amount of $CuSO_4$ = D mol
Molar mass of $CuSO_4 \cdot nH_2O$ = (143.6 + 18n) g mol^{-1}
Mass of $CuSO_4 \cdot nH_2O$ = {D × (143.6 + 18n)} g
But mass of $CuSO_4 \cdot nH_2O$ = (Y − X) = {D × (143.6 + 18n)} ⇒ solve for n!

 Q Can you roughly show us how to calculate the relative atomic mass of **M** in MCO_3?

A: $$MCO_3(s) \rightarrow MO(s) + CO_2(g)$$

Loss in mass gives the number of moles of $CO_2 = b$ mol
But number of moles of CO_2 = number of moles of MCO_3

$$= \frac{\text{Mass of } MCO_3 \text{ used for heating}}{\text{Relative molecular mass of } MCO_3} = b$$

$$\Rightarrow \frac{\text{Mass of } MCO_3 \text{ used for heating}}{\text{Relative atomic mass of } M + 12.0 + 3(16.0)} = b.$$

Hence, if both the mass of MCO_3 and the number of moles of CO_2 are known, the relative atomic mass of **M** can be easily determined.

> **Q** So, the whole idea is: through the measurement of the mass, we determine the number of moles of a particular substance and then with the help of a chemical equation, we obtain the number of moles of a reactant?

A: It is great that you can see the guiding principles behind the planning exercise. For titration planning, it is through the measurements of volume and concentration that we determine the number of moles. But similarly, we need the help of a chemical equation in order to process our data.

2.2 Mixture of Lead Carbonate and Barium Carbonate

The Task:

A sample of lead carbonate is contaminated with some barium carbonate. You are supposed to plan an experiment, using a precipitation reaction to determine the actual mass of lead carbonate in the sample with the following chemicals and apparatus:

- 1.2 g of lead carbonate with barium carbonate;
- Dilute nitric acid;
- Dilute hydrochloric acid;
- Dilute sulfuric acid;
- Crucible; and
- Oven.

> **Q** Where should we start thinking?

A: (i) What is the purpose of the plan?
— To determine the mass of lead carbonate in the mixture of lead and barium carbonates.
(ii) What do you need to know in order to know the mass of lead carbonate?
— We need to know the number of moles of lead carbonate in a fixed mass of the sample that we have measured.
(iii) How do you then determine the number of moles of lead carbonate?
— We can heat the sample to decompose the lead carbonate. But wait a minute, the barium carbonate may also decompose. Well, we can

add nitric acid to dissolve the sample and then add sulfuric acid to precipitate out the PbSO$_4$. Hey man! But the BaSO$_4$ would also be precipitated out. So, we can only use hydrochloric acid to dissolve the mass sample as BaCl$_2$ will not be precipitated out:

$$Pb^{2+}(aq) + 2Cl^-(aq) \rightarrow PbCl_2(s).$$

Thus, if we know the mass of PbCl$_2$ that is precipitated out, we can calculate its number of moles and then the number of moles of Pb^{2+} ions that comes from the PbCO$_3$ in the sample.

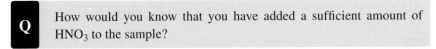

Q: How would you know that you have added a sufficient amount of HNO$_3$ to the sample?

A: When acid is added to a carbonate, there would be effervescence. All the carbonate would have reacted with acid if a clear solution is formed and there is no further effervescence observed. Otherwise, you can do a pre-calculation to find out the volume of acid that is needed if the concentration of the acid is given.

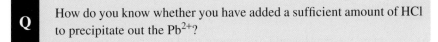

Q: How do you know whether you have added a sufficient amount of HCl to precipitate out the Pb^{2+}?

A: After the precipitate is formed, let the suspension settle first. Then, add drops of HCl solution and observe whether there is further formation of precipitate at the surface when the droplets of HCl solution come in contact with the surface of the solution. Otherwise, you can do a pre-calculation to find out the volume of acid that is needed if the concentration of the acid is given. Thereafter, simply add excess acid to precipitate out all the Pb^{2+} ions.

The Procedure:

Procedure for the heating:

(i) *Weigh* accurately *0.6 g* of the sample into a small *beaker*.
(ii) Use a *measuring cylinder*, add 20 cm^3 of HNO$_3$ solution. *Stir* well and ensure *no further effervescence* is observed.
(iii) Use another *measuring cylinder*, add *20 cm^3* of HCl solution to the resulting solution in step (ii). *Stir* well to mix.

(iv) Let the *suspension settle down*. Use a dropper to add *drops of* HCl *solution*. Observe whether there is *any further precipitation*. If yes, then continue to add HCl solution till no further precipitation is observed.
(v) *Filter* the mixture with a *pre-weighed crucible*.
(vi) *Wash* the residue with some *cold water*.
(vii) *Dry* the crucible with residue in an *oven*.
(viii) Let the crucible with residue *cool down in a dessicator*.
(ix) *Weigh* the crucible with the residue till *constant mass. Calculate* the mass of *residue* obtained.
(x) *Repeat the experiment* with another sample.

Q Why do you need to wash the residue with cold water?

A: The washing is to get rid of any solution that contains undesired ions that may cling onto the surface of the solid. As to why cold water is used — it does not dissolve as much of the solid as water at room temperature would.

Q Does this mean that we cannot wash the residue too many times with water?

A: Yes, more washings would cause more solid to dissolve. So, there must be a balance between getting a purer compound and the desired quantity.

Q How can we increase the reliability of the experiment?

A: Well, you can repeat the drying, cooling, and weighing process till you get a constant mass.

Q In what other precipitation reactions can the above planning outline be applied to?

A: Many! For example:
— Determine the value of n of $CuSO_4 \cdot nH_2O$:
You can determine the amount of SO_4^{2-} by precipitating out $BaSO_4$ or $PbSO_4$. But if there are contaminants such as Cl^-, then you cannot use $Pb(NO_3)_2$ as the insoluble $PbCl_2$ would also be precipitated out.

- Determine the purity of BaO_2:
  You can determine the amount of BaO_2 by precipitating out $BaSO_4$; hydrogen peroxide would be formed too. But if there are contaminants such as Pb^{2+}, then you cannot use H_2SO_4 as the insoluble $PbSO_4$ would also be precipitated out:
  $$BaO_2(s) + H_2SO_4(aq) \rightarrow BaSO_4(s) + H_2O_2(aq).$$
- Determine the identity of a sulfate MSO_4:
  You can determine the amount of SO_4^{2-} by precipitating out $BaSO_4$ or $PbSO_4$. But if there are contaminants such as Cl^-, then you cannot use $Pb(NO_3)_2$ as the insoluble $PbCl_2$ would also be precipitated out.
- Determine the identity of the metal cation in MCl:
  You can determine the amount of Cl^- by precipitating out $AgCl$ or $PbCl_2$. But if there are contaminants such as SO_4^{2-}, then you cannot use $Pb(NO_3)_2$ as the insoluble $PbSO_4$ would also be precipitated out.
- Determine the amount of Zn^{2+} in a sample:
  You can determine the amount of Zn^{2+} by precipitating out $ZnCO_3$ using aqueous Na_2CO_3. You cannot use $NaOH$ to precipitate $Zn(OH)_2$ out because the $Zn(OH)_2$ is soluble in excess $NaOH$. So, the result may be highly inaccurate.

It would be good if you could remember some of the soluble and insoluble salts:

Soluble Salts	Insoluble Salts
• Salts containing Na^+, K^+ and NH_4^+ as cations are soluble.	
• Salts containing NO_3^- as the anion are soluble.	
• Salts containing Cl^-, Br^- and I^- as the anion are soluble EXCEPT	• $PbCl_2$, $PbBr_2$, PbI_2. • $AgCl$, $AgBr$, AgI
• Salts containing SO_4^{2-} as the anion are soluble EXCEPT	• $PbSO_4$, $BaSO_4$ and $CaSO_4$ (sparingly soluble)
• Na_2CO_3, K_2CO_3, $(NH_4)_2CO_3$.	• Salts containing CO_3^{2-} as the anion are insoluble EXCEPT

Do you notice some of the above experiments have been discussed using titration planning in Chapter 1?

Q Cu^{2+} reacts with I^- according to the equation, $2Cu^{2+}(aq) + 4I^-(aq) \rightarrow 2CuI(s) + I_2(aq)$. Now, since CuI is insoluble in water, can we use gravimetric analysis to determine the value of n of $CuSO_4 \cdot nH_2O$ by weighing the CuI precipitate?

A: It is not encouraged because if too much iodine is formed, the iodine will also be crystallized. Thus, you would filter the insoluble iodine crystals and CuI together.

Q Can we precipitate out Fe^{2+} as $Fe(OH)_2$?

A: No! This is because the $Fe(OH)_2$ is highly unstable and undergoes oxidation to form $Fe(OH)_3$. Thus, the mass measurement is highly inaccurate. It would be good to oxidize the Fe^{2+} to Fe^{3+} first, then precipitate it out as $Fe(OH)_3$.

Q Can we precipitate the Fe^{3+} out as $Fe_2(CO_3)_3$?

A: Have you forgotten that Fe^{3+} in water would give us an acidic solution that is strong enough to decompose CO_3^{2-} ions? For more details, please refer to *Understanding Advanced Physical Inorganic Chemistry* by J. Tan and K. S. Chan.

2.3 A Sample of Impure Sodium Thiosulfate

The Task:

Thiosulfate ion, $S_2O_3^{2-}$, decomposes in the presence of acid to give solid sulfur and sulfur dioxide gas:

$$S_2O_3^{2-}(aq) + 2H^+(aq) \rightarrow S(s) + SO_2(g) + H_2O(l).$$

A student obtained some impure sodium thiosulfate. Describe how you would help the student to determine the percentage of sodium thiosulfate in the impure sample. You are provided with the following chemicals and apparatus:

- 2.00 mol dm^{-3} HCl solution;
- 10 g of impure sodium thiosulfate;
- Crucible; and
- Oven.

Q	Where should we start thinking?

A: (i) What is the purpose of the plan?
- To determine the percentage of $Na_2S_2O_3$ in a sample of impure sodium thiosulfate.

(ii) What do you need to know in order to find the percentage of $Na_2S_2O_3$?
- We need to know the number of moles of $Na_2S_2O_3$ in a fixed mass of the sample that we have measured.

(iii) How can you then determine the number of moles?
- We can make use of the following balanced equation for the reaction between $Na_2S_2O_3$ and HCl:

$$S_2O_3^{2-}(aq) + 2H^+(aq) \rightarrow S(s) + SO_2(g) + H_2O(l).$$

Thus, if we know the number of moles of sulfur that has been collected, we would be able to deduce the number of moles of $Na_2S_2O_3$ that is present in the sample.

(iv) But can you add the acid to the solid $Na_2S_2O_3$ directly?
- No, we need to dissolve it in water first.

(v) Do you know how much HCl to add?
- Well, we can assume that a certain mass of $Na_2S_2O_3$ is used and then calculate the theoretical amount of HCl that is needed.

Pre-Experimental Calculations:

$$S_2O_3^{2-}(aq) + 2H^+(aq) \rightarrow S(s) + SO_2(g) + H_2O(l)$$

Assuming that we have used about 5 g of impure $Na_2S_2O_3$ and all the 5 g is due to $Na_2S_2O_3$ only:

Molar mass of $Na_2S_2O_3$ = $2(23.0) + 2(32.1) + 3(16.0) = 158.2$ g mol^{-1}

Amount of $Na_2S_2O_3$ in 5 g = $\frac{5}{158.2} = 3.16 \times 10^{-2}$ mol

Amount of HCl needed = $2 \times$ Amount of $Na_2S_2O_3$ in 5 g
= $2 \times 3.16 \times 10^{-2} = 6.32 \times 10^{-2}$ mol

Volume of 2.00 mol dm^{-3} HCl solution needed = $\frac{6.32 \times 10^{-2}}{2} = 0.0316$ dm^3
= 31.6 cm^3.

The Procedure:

Procedure for precipitating out the insoluble sulfur:

(i) *Weigh accurately 5.0 g* of the impure in a clean dry *beaker*.
(ii) Use a *measuring cylinder*, add in *50 cm^3* of water to dissolve it. If the solid does not dissolve, add more water to dissolve it.
(iii) Use a *measuring cylinder*, add in *33 cm^3* of 2.00 mol dm^{-3} HCl solution.
(iv) *Let the sulfur suspension settle down.* Use a *dropper* to add *drops of HCl solution*. Observe whether there is *further precipitation*. If yes, then continue to add HCl solution till no further precipitation is observed.
(v) *Filter* the mixture with a *pre-weighed crucible*.
(vi) *Wash* the residue with some *cold water*.
(vii) *Dry* the crucible with residue in an *oven*.
(viii) Let the crucible with residue *cool down in a dessicator*.
(ix) *Weigh* the crucible with the residue till *constant mass*. *Calculate* the mass of *residue* obtained.
(x) *Repeat the experiment* with another sample.

Q Why must we ensure that all the sample has dissolved before adding the acid?

A: If the sample does not all dissolve before the addition of the acid, the sulfur that is precipitated out may clump together with the unreacted sodium thiosulfate. This would prevent the unreacted sodium thiosulfate from further reactions. The result would be inaccurate.

2.4 Determine the Solubility of a Solid/Determine the K_{sp} Value of a Partially Soluble Compound

The Task:

Solubility can be defined as the mass of a substance that dissolved per 100 g of water at a particular temperature. You are given some solid **X**. Describe how you would determine the solubility, in grams per 100 g of water, of solid **X** at temperature **T**°C. You are provided with the following chemicals and apparatus:

- Solid **X**;
- Deionized water;
- 10.0 cm^3 pipette;
- Thermostated water bath; and
- Evaporating dish.

Q Where should we start thinking?

A: (i) What is the purpose of the plan?
— To determine the solubility, in grams per 100 g of water, of solid **X** at temperature **T**°C.

(ii) What do you need to know in order to determine the solubility, in grams per 100 g of water, of solid **X** at temperature **T**°C?
— We need to know the mass of solid that has dissolved in a specific volume of water.

(iii) How do you find out the mass of the solid that has dissolved?
— We can use the pipette to draw out an accurate volume of the filtrate, weigh it and then evaporate the solution to dryness. Thus, knowing the mass of water and the mass of the dissolved solute, we can calculate the solubility, in grams per 100 g of water, of solid **X** at temperature **T**°C.

(iv) But how do you ensure that the solubility is measured at **T**°C?
— Well, the thermostated water bath will help us to maintain the desired temperature.

The Procedure:

(i) Use a *measuring cylinder*, measure *50 cm^3* of deionized water and introduce it into a *beaker* place in the *thermostated water bath at* **T**°C.
(ii) Add solid **X** with *stirring* until solid **X** *does not dissolve anymore*.
(iii) *Filter* the mixture into a *dry and clean conical flask at* **T**°C, so that the filtrate that is collected is still at **T**°C.

(iv) Use a *clean and dry pipette* to draw out *10.0 cm³* of the filtrate and placed it in a *pre-weighed, clean and dry evaporating dish* (m_1).
(v) *Weigh* the evaporating dish and its content (m_2).
(vi) *Heat, cool and then weigh* the evaporating dish until a *constant mass* is obtained (m_3).
(vii) *Repeat the experiment* with another sample.
(viii) The solubility of solid **X** per 100 g of water = $\frac{m_3 - m_1}{m_2 - m_3} \times 100$.

Q If solid **X** decomposes on heating, can we still use the above method?

A: Of course not.

Q Is there any unit or dimension for the above solubility value?

A: Yes, of course. It is grams per 100 g of water or g (100 g of water)$^{-1}$. It is not a dimensionless quantity.

Q If the solubility is defined as the mass of solid dissolved per 100 cm³ of water, how would the procedure be different?

A: The Procedure:
(i) Use a 50 cm³ *burette* to measure *50 cm³* of deionized water and introduce it into a *clean and dry beaker* place in the *thermostated water bath* at **T°C**.
(ii) Add an *accurate mass of solid* **X** (m_1) with *stirring* until solid **X** *does not dissolve anymore*.
(iii) *Filter* the mixture using a *clean and dry pre-weighed crucible maintained* at **T°C**.
(iv) *Dry* the crucible with residue in an *oven*.
(v) Let the crucible with residue *cool down in a dessicator*.
(vi) *Weigh* the crucible with the residue till *constant mass*. Calculate the mass of *residue* obtained (m_2).
(vii) *Repeat the experiment* with another sample.
(viii) The solubility of solid **X** per 100 cm³ of water = $\frac{m_1 - m_2}{50} \times 100$.

Q So, if the question wants us to plot a graph of solubility against temperature, all we need to do is to just follow the above procedure but change the **T°C** values?

A: Yes, you are right. Refer to the procedure below:

The Procedure:

(i) Use a *measuring cylinder*, measure *50 cm^3* of deionized water and introduce it into a *beaker* place in the *thermostated water bath at* **T°C**.
(ii) Add solid **X** with *stirring* until solid **X** *does not dissolve anymore*.
(iii) *Filter* the mixture into a *dry and clean conical flask at* **T°C**, so that the filtrate that is collected is still at **T°C**.
(iv) Use a *clean and dry pipette* to draw out *10.0 cm^3* of the filtrate and placed it in a *pre-weighed, clean and dry evaporating dish* (m_1).
(v) *Weigh* the evaporating dish and its content (m_2).
(vi) *Heat, cool, and then weigh* the evaporating dish until a *constant mass* is obtained (m_3).
(vii) *Repeat the experiment* with another sample at a different **T°C** value.
(viii) The solubility of solid **X** per 100 g of water = $\frac{m_3 - m_1}{m_2 - m_3} \times 100$.
(ix) Plot a graph of solubility against **T°C** value.

The plan is also applicable for the following task. The main difference is that instead of changing the temperature, **T°C**, you just have different concentration of H_2SO_4.

The Task:

You are given some pure cerium(IV) sulfate(VI), $Ce(SO_4)_2$. Plan a gravimetric experiment to determine how the solubility of cerium(IV) sulfate(VI) in dilute sulfuric(VI) acid would depend on the concentration of the acid. The procedure is as follows:

The Procedure:

(i) Use a 50 cm^3 *burette* to measure *50 cm^3* of *sulfuric acid* and introduce it into a *clean and dry beaker* place in the *thermostated water bath at* **T°C**.

(ii) Add in an *accurate mass of solid* **X** (m_1) with *stirring* until solid **X** does not dissolve anymore.
(iii) *Filter* the mixture using a *clean and dry pre-weighed crucible* maintained at **T°C**.
(iv) *Wash* the residue with some *cold water*.
(v) *Dry* the crucible with residue in an *oven*.
(vi) Let the crucible with residue *cool down in a dessicator*.
(vii) *Weigh* the crucible with the residue till *constant mass. Calculate the mass of residue* obtained (m_2).
(viii) *Repeat the experiment* with another sample but using sulfuric acid of different concentration values.
(ix) The solubility of solid **X** per 100 cm^3 of water = $\frac{m_1 - m_2}{50} \times 100$.
(x) *Plot a graph* of solubility against concentration of sulfuric acid.

Q Can the following task use the above planning approach?

The Task:
You are given some pure barium carbonate, $BaCO_3$. Plan a gravimetric experiment to determine whether the presence of carbon dioxide in water does increase the solubility of $BaCO_3$.

A: Certainly! The titration approach has been discussed in Section 1.6 of Chapter 1. All you need to do is dissolve some known mass of $BaCO_3$ in both deionized water and deionized water saturated with CO_2. Then, perform a filtration and measure the mass of the residue collected for the two different types of water.

Q After the experiment, how can we make use of the solubility data for $BaCO_3$?

A: Good question! You can use it to calculate the solubility product constant for $BaCO_3$:

$$BaCO_3(s) \rightleftharpoons Ba^{2+}(aq) + CO_3^{2-}(aq), \quad K_{sp} = [Ba^{2+}(aq)][CO_3^{2-}(aq)].$$

2.5 Determine the Concentration of an Acid through Gravimetry

The Task:

You are given some hydrochloric acid which may have a concentration of either 1 mol dm^{-3} or 2 mol dm^{-3}. Describe how you would determine the concentration of the acid through gravimetry. You are provided with the following chemicals and apparatus:

- HCl of unknown concentration;
- 1 mol dm^{-3} AgNO$_3$ solution;
- Weighing balance;
- Crucible; and
- Oven.

Q Where should we start thinking?

A: (i) What is the purpose of the plan?
— To determine the concentration of the acid.
(ii) What do you need to know in order to determine the concentration of the acid?
— We need to know the amount of Cl$^-$ ion per unit volume of the acid.
(iii) How do you then determine the amount of Cl$^-$ ion per unit volume?
— There are two possible concentrations, either 1 mol dm^{-3} or 2 mol dm^{-3}, and the concentration of AgNO$_3$ is also 1 mol dm^{-3}. If you mix 10 cm^3 of AgNO$_3$ and 20 cm^3 of AgNO$_3$, each with 10 cm^3 of the acid, the mixture that gives you a greater amount of AgCl precipitate must be the one with a concentration of 2 mol dm^{-3}. But if both mixtures give you the same mass of AgCl, then the concentration of the acid is 1 mol dm^{-3}.

The Procedure:

(i) Use a *measuring cylinder*, each measure *10 cm^3* of the acid and place it into two separate *clean and dry beakers*.

(ii) Use another *measuring cylinder*, add *10* cm^3 of AgNO$_3$ into beaker 1 and add *20* cm^3 of AgNO$_3$ into beaker 2.
(iii) *Filter* the two mixtures with two *pre-weighed crucibles*.
(iv) *Wash* the two residues with some *cold water*.
(v) *Dry* the crucibles with residue in an *oven*.
(vi) Let the crucibles with residue *cool down in a dessicator*.
(vii) *Weigh* the crucibles with the residue till *constant mass*. Calculate the masses of the two *residues* obtained.
(viii) If the two residues have about the same mass, then the acid is of concentration 1 mol dm^{-3}. But if one residue has a mass that is almost doubled that of the other, the concentration of the acid is 2 mol dm^{-3}.

Q

Can the above plan be used for the following task?

The Task:

You are given some silver nitrate which may have a concentration of either 1 mol dm^{-3} or 2 mol dm^{-3}. Describe how you would determine the concentration of the silver nitrate solution through gravimetry. You are provided with the following chemicals and apparatus:

- Silver nitrate solution of unknown concentration;
- 1 mol dm^{-3} HCl solution;
- Crucible; and
- Oven.

A: Certainly.

2.6 Safety Precautions for Gravimetry

You may be asked to quote some safety precautions while performing a gravimetric experiment. Depending on the type of gravimetric experiment that you are performing, the following examples may be useful for you to take note:

— If thermal decomposition gives off a toxic gas such as nitrogen dioxide, NO_2, perform the experiment in a fumehood.
— Always wear gloves, a lab coat, and safety goggles while doing the experiment. For example: calcium oxide, CaO, causes irritation to the

skin and eye. Hence, there should be minimal direct contact with these chemicals.
— After heating, it is good to allow the crucible/boiling tube to cool sufficiently before handling it. Use a pair of tongs/test-tube holder to handle hot objects.
— While heating a boiling tube, point it away from yourself and others. This is to ensure that no one is injured when the contents spurt during heating.
— While heating a sample using a boiling tube, move the boiling tube up and down, and do not heat the tube at just one spot. This is to prevent the contents of the boiling tube from spurting due to intensive heating at one spot.
— Clamp your filtration flask with a retort clamp, especially during filtration if you are using a Buchner flask (refer to Chapter 9). This is to prevent the filtration flask from accidentally toppled over.

2.7 Minimizing Experimental Errors or Increasing Reliability

You may be asked to quote some ways to improve experimental reliability while performing a gravimetric experiment. Depending on the type of gravimetric experiment that you are performing, the following examples may be useful for you to take note:

— Heat the solid sample evenly to ensure all the solid has decomposed.
— Cool the heated sample in a dessicator to prevent the absorption of moisture in the air.
— Repeat the heating–cooling–weighing cycle till constant mass is achieved.
— Ensure all ions are precipitated out during a precipitation experiment before filtration. This can be done by using excess precipitating reagent.
— Wash the filtered residue with a suitable solvent that would remove impurities well, but at the same time, does not dissolve the residue.

CHAPTER 3

PLANNING USING THE GAS COLLECTION METHOD

For chemical reactions that give off a gaseous product, by measuring the volume of the gas evolved when an excess of reagent is added to the analytes, quantitative information may be obtained to determine the amount of the analyte that is present. The gas evolved may be collected using a graduated frictionless syringe or simply by downward displacement of water, using an inverted measuring cylinder or burette. Certainly, the gas that is collected needs to have minimal solubility in water, else it would affect the accuracy of the result.

The experimental set-ups for gas collection can be as follows:

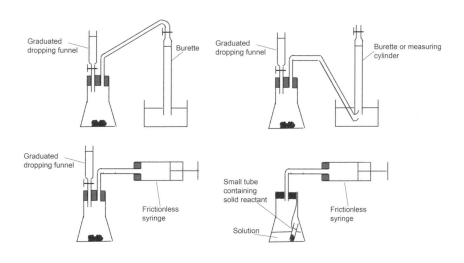

Q What is the purpose of the graduated dropping funnel?

A: The graduated dropping funnel is to introduce the liquid reagent without loss of any gas produced the moment the liquid reagent comes in contact with the other reagent that is already inside the flask. Excess reagent is left in the graduated dropping funnel after discharging the reagent to minimize the loss of gas when the tap is opened.

Q What precaution do we need to take note when we use the graduated dropping funnel set-up?

A: Take note that the volume of solution that you have introduced into the flask would displace the same volume of air, which would then be collected by the syringe. So, you need to subtract the volume of the solution that is introduced from the volume of the gas that is collected.

Q Why do we need a frictionless syringe?

A: If the syringe is not frictionless, you might not be able to measure the volume of the gas accurately as the friction between the plunger and the syringe would prevent the plunger from moving. This would cause the pressure to build up inside the syringe.

Q What is the advantage of collecting the gas using a syringe as compared to collecting the gas using the downward displacement of water?

A: If the gas is soluble in water, collection of the gas using the syringe is more accurate than collecting the gas using the downward displacement of water. The table below shows us the relative solubilities of some gases in water:

Insoluble	Moderate Solubility	Very Soluble
CO_2, H_2, O_2, N_2, CO	SO_2, Cl_2	NH_3, HCl, HBr, HI

Take note that even though a gas may be classified as insoluble in water, it still has some degree of solubility. Hence, the collection of the gas via a gas syringe is preferred. Also, the syringe must be frictionless or well greased.

In addition, the volume of the gas that is measured using the downward displacement of water also includes the volume of the water vapor. Hence, to calculate the actual volume of the gas that is collected, one needs to subtract the volume attributed to the water vapor.

Q Why are we able to determine the quantity of the analyte by measuring the volume of gas collected?

A: From the ideal gas equation, $pV = nRT$, under fixed pressure and temperature, the volume (V) of a gas is directly proportional to its number of moles (n). There are two molar volumes you need to remember well:

At s.t.p. (1 bar, 0°C), molar volume = 22.7 dm^3 mol^{-1}; and
At r.t.p. (1 atm, 20°C), molar volume = 24.0 dm^3 mol^{-1}.

Q Is 1 bar equivalent to 1 atm?

A: No! 1 bar is 100 kPa whereas 1 atm is 101.325 kPa. So, there is a slight difference.

Q What is the preferred volume of gas to be collected during an experiment?

A: Well, 50 cm^3 is a good number. Too small a volume collected would incur too much experimental error. Too large a volume collected would need more sample and reagents and takes more time.

3.1 Mixture of Potassium Hydrogen Carbonate and Potassium Chloride

The Task:

A sample of potassium hydrogen carbonate, KHCO$_3$, contains 4–8% of potassium chloride, KCl. You are supposed to plan an experiment using gas collection to determine the actual percentage of the potassium hydrogen carbonate in the sample with the following chemicals and apparatus:

- 0.100 mol dm^{-3} H$_2$SO$_4$ solution;
- 1 g sample of potassium hydrogen carbonate and potassium chloride mixture;
- Frictionless gas syringe; and
- Standard glassware in the lab.

Q Where should we start thinking?

A: (i) What is the purpose of the plan?
— To determine the percentage by mass of the potassium hydrogen carbonate in the mixture.

(ii) What do you need to know in order to determine the mass of potassium hydrogen carbonate?
— We need to know the number of moles of potassium hydrogen carbonate in a fixed mass of the sample that we have measured.

(iii) How can you then determine the number of moles of potassium hydrogen carbonate?
— We can make use of the following balanced equation for the reaction between KHCO$_3$ and H$_2$SO$_4$:

$$2KHCO_3(s) + H_2SO_4(aq) \rightarrow K_2SO_4(aq) + 2CO_2(g) + 2H_2O(l).$$

Thus, if we know the number of moles of CO$_2$ gas collected, we would be able to deduce the number of moles of KHCO$_3$ present in the sample.

(iv) How do you determine the volume of H$_2$SO$_4$ solution that is used?
— We can assume that the sample is 100% pure, consisting only of KHCO$_3$. From here, we can calculate the theoretical amount of KHCO$_3$ that is present. From the above equation, we can proceed to

determine the number of moles of H_2SO_4 that is needed and hence, its volume that is needed.

(v) But how would you know how much of the sample to weigh?
— We can assume that we need to collect 50 cm³ of the gas and from here determine the number of moles of CO_2 and then the number of moles of $KHCO_3$ from the above balanced equation.

Pre-Experimental Calculations:

$$2KHCO_3(aq) + H_2SO_4(aq) \rightarrow K_2SO_4(aq) + 2CO_2(g) + 2H_2O(l)$$

Assuming that 50 cm³ of CO_2 is collected at r.t.p. conditions:
At r.t.p. (1 atm, 20°C), molar volume = 24.0 dm³ mol⁻¹
Amount of CO_2 gas = $\frac{50}{24,000}$ = 2.08 × 10⁻³ mol
Amount of $KHCO_3$ present = Amount of CO_2 gas = 2.08 × 10⁻³ mol
Assuming that the sample consists only of $KHCO_3$:
Molar mass of $KHCO_3$ = 39.1 + 1.0 + 12.0 + 3(16.0) = 100.1 g mol⁻¹
Mass of $KHCO_3$ present = 2.08 × 10⁻³ × 100.1 = 0.2085 g
Amount of H_2SO_4 used = $\frac{1}{2}$ × Amount of CO_2 gas = 1.04 × 10⁻³ mol
Volume of 0.100 mol dm⁻³ H_2SO_4 solution needed = $\frac{1.04 \times 10^{-3}}{0.100}$ = 0.0104 dm³
= 10.4 cm³.

The Procedure:

The general procedure for gas collection experiments involves measuring the volume of the gas that is generated via a chemical reaction at a specific temperature and pressure. Thus, the procedure for a typical gas collection experiment consists of a series of steps which inform the student: (1) which step should come first; (2) what apparatus should he/she use; (3) what is the quantity of substance that he/she should measure; and (4) if needed, the reaction conditions such as temperature and pressure.

Procedure for gas collection:

(i) Set up the apparatus as shown (refer to p. 77).
(ii) Accurately measure *0.20 g* of the sample and place it in the *conical flask*.
(iii) Record the *initial reading of the syringe*.
(iv) Using the *graduated dropping funnel*, introduce *12.0 cm^3* of the solution into the conical flask. *Swirl* to mix *gently*.
(v) Allow the reaction to progress *until the plunger of the syringe does not move*.
(vi) *Allow time for the volume of the gas to equilibrate to the temperature and pressure*.
(vii) Record the *final reading of the syringe*. Remember to *subtract the volume of the solution that is introduced from the final reading of the syringe*.
(viii) *Repeat the experiment* to check for *reliability of results*.

Q   Why do we need to allow time for the temperature and pressure to equilibrate before taking the final reading of the syringe?

A: As the reaction goes, heat may be evolved. The temperature and pressure of the gas collected may not be equal to the external temperature and pressure. In addition, gas "moves" because of pressure difference. Thus, at this point, if you take the reading immediately, there is a high degree of inaccuracy.

Q   If we need to collect the gas dry, is there a way?

A: If the gas is neither basic nor acidic, you can dry it by passing through drying agents such as concentrated sulfuric(VI) acid, anhydrous calcium chloride, anhydrous copper(II) sulfate, calcium(II) oxide, and silica gel. But if the gas is an acidic gas, you cannot use calcium(II) oxide as the latter is a base. And if it is basic, do not use concentrated sulfuric(VI) acid. For ammonia gas, calcium chloride cannot be used as the drying agent because of the following reaction that forms an additional compound:

$$CaCl_2(s) + 8NH_3(g) \rightarrow CaCl_2 \cdot 8NH_3(s).$$

Isn't the ammonia in $CaCl_2 \cdot 8NH_3$ similar to the "water of crystallization" in $CuSO_4 \cdot 5H_2O(s)$?

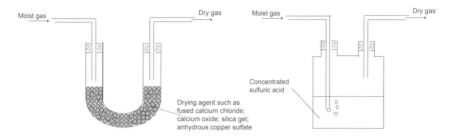

3.2 A Sample of Insoluble $BaCO_3$

The Task:

A student obtained some impure barium carbonate, $BaCO_3$, during a field trip. Describe how you would help the student to determine the percentage of $BaCO_3$ in the impure sample by the gas collection method. You are provided with the following chemicals and apparatus:

- 0.100 mol dm^{-3} HCl solution;
- 1.0 g of impure $BaCO_3$;
- Frictionless gas syringe; and
- Standard glassware in the lab.

Q Where should we start thinking?

A: (i) What is the purpose of the plan?
— To determine the percentage of $BaCO_3$ in the impure sample.
(ii) What do you need to know in order to determine the percentage of $BaCO_3$?
— We need to know the number of moles of $BaCO_3$ in a fixed mass of the sample that we have measured.

(iii) How do you then determine the number of moles of $BaCO_3$?
— We can make use of the following balanced equation for the reaction between $BaCO_3$ and HCl:

$$BaCO_3(s) + 2HCl(aq) \rightarrow BaCl_2(aq) + CO_2(g) + H_2O(l).$$

Thus, if we know the number of moles of CO_2 gas collected, we would be able to deduce the number of moles of $BaCO_3$ present in the sample.

(iv) How do you determine the volume of HCl solution that is used?
— We can assume that the sample is 100% pure, consisting only of $BaCO_3$. From here, we can calculate the theoretical amount of $BaCO_3$ present. From the above equation, we can proceed to determine the number of moles of HCl needed.

(v) But how would you know how much of the sample to weigh?
— We can assume that we need to collect 50 cm^3 of the gas and from here determine the number of moles of CO_2 and then the number of moles of $BaCO_3$ from the above balanced equation.

Pre-Experimental Calculations:

$$BaCO_3(s) + 2HCl(aq) \rightarrow BaCl_2(aq) + CO_2(g) + H_2O(l)$$

Assuming that 50 cm^3 of CO_2 gas is collected at r.t.p. conditions:
At r.t.p. (1 atm, 20°C), molar volume = 24.0 dm^3 mol^{-1}
Amount of CO_2 gas = $\frac{50}{24,000}$ = 2.08 × 10^{-3} mol
Amount of $BaCO_3$ present = Amount of CO_2 gas = 2.08 × 10^{-3} mol
Assuming the sample consists only of $BaCO_3$:
Molar mass of $BaCO_3$ = 137.3 + 12.0 + 3(16.0) = 197.3 g mol^{-1}
Mass of $BaCO_3$ present = 2.08 × 10^{-3} × 197.3 = 0.4104 g
Amount of HCl used = 2 × Amount of CO_2 gas = 4.16 × 10^{-3} mol
Volume of 0.100 mol dm^{-3} HCl solution needed = $\frac{4.16 \times 10^{-3}}{0.100}$ = 0.0416 dm^3
= 4.16 cm^3.

Q It seems that for during the planning for a gas collection experiment, the pre-calculation always start off by assuming that a certain volume of gas is collected. Thereafter, we would determine the amounts of reagents that are needed for the experiment.

A: Absolutely spot on!

 Q Are there any other assumptions made other than assuming that there is only $BaCO_3$ in the impure sample?

A: Well, we have to also assume that the impurities present in the impure sample do not react with the acid used.

The Procedure:

Procedure for gas collection:
  (i) Set up the apparatus as shown (refer to p. 77).
  (ii) Accurately measure *0.40 g* of the sample and introduce it into the *conical flask*.
  (iii) Record the *initial reading of the syringe* (V_1).
  (iv) Using the *graduated dropping funnel*, run in *42.0 cm³* of the HCl solution. *Swirl* to mix *gently*.
  (v) Allow the reaction to progress *until the plunger of the syringe does not move*.
  (vi) *Allow time for the volume of the gas to equilibrate to the temperature and pressure.*
  (vii) Record the *final reading of the syringe* (V_2). Remember to *subtract the volume of solution that is introduced from the final reading of the syringe*.
  (viii) *Repeat the experiment* to check for *reliability of results*.

 Q Can we use sulfuric(VI) acid to replace HCl?

A: No, the SO_4^{2-} ion will form an insoluble precipitate and coat around $BaCO_3$, preventing further reaction. But for HCl, the $BaCl_2$ that is formed is soluble.

Results and Calculations:

Record the volume measurements in the table as shown below:

Initial gas syringe reading/cm^3	V_1
Final gas syringe reading/cm^3	V_2
Volume of gas evolved/cm^3	$V_2 - V_1$ (if the reagent was NOT introduced via the graduated dropping funnel)
	$V_2 - V - V_1$ (if the reagent, V cm^3, was introduced via the graduated dropping funnel)

Let the volume of the gas that is collected be V cm^3.

$$BaCO_3(s) + 2HCl(aq) \rightarrow BaCl_2(aq) + CO_2(g) + H_2O(l)$$

At r.t.p. (1 atm, 20°C), molar volume = 24.0 dm^3 mol^{-1}

Amount of CO_2 gas = $\dfrac{V}{24,000}$ mol

Amount of $BaCO_3$ present = Amount of CO_2 gas = $\dfrac{V}{24,000}$ mol

Molar mass of $BaCO_3$ = 137.3 + 12.0 + 3(16.0) = 197.3 g mol^{-1}

Mass of $BaCO_3$ present = $\dfrac{V}{24,000} \times 197.3 = \dfrac{197.3V}{24,000}$ g.

Note: In addition, such a planning process can also be used for the following similar cases:

(i) Determine the purity of $PbSO_3$ using nitric acid:

$$PbSO_3(s) + 2HNO_3(aq) \rightarrow Pb(NO_3)_2(aq) + SO_2(g) + H_2O(l).$$

(Take note that both H_2SO_4 and HCl cannot be used because of insoluble $PbSO_4$ and $PbCl_2$ that will form, respectively. In addition, due to the relatively good solubility of SO_2 in water, not all the SO_2 gas evolved can be accurately collected. So, the latter would decrease the accuracy of the experimental result if downward displacement of water is used.)

(ii) Determine the relative atomic mass of lithium (or any metal, such as Zn, Mg, Fe, etc., that would react with the acid):

$$2Li(s) + 2HCl(aq) \rightarrow 2LiCl(aq) + H_2(g).$$

(Take note that if the metal is too reactive, the weaker acid such as water should be used.)

(iii) Determine the relative atomic mass of the element **M** in the metal carbonate, **MCO$_3$** or **M$_2$CO$_3$**:

$$MCO_3(s) + 2HCl(aq) \rightarrow MCl_2(aq) + CO_2(g) + H_2O(l);$$

$$M_2CO_3(s) + 2HCl(aq) \rightarrow 2MCl(aq) + CO_2(g) + H_2O(l).$$

The main idea for all these planning exercises is to collect a particular volume of the gas, calculate its number of moles based on the collected volume, and then determine the number of moles of the analyte using a balanced equation.

3.3 Determination of the Ideal Gas Constant, R, Using BaCO$_3$

The Task:

A student wants to determine the ideal gas constant, R, of the ideal gas equation, $pV = nRT$ by using the following reaction:

$$BaCO_3(s) + 2HCl(aq) \rightarrow BaCl_2(aq) + CO_2(g) + H_2O(l).$$

Describe how you would help the student to determine the value of R by the gas collection method. You are provided with the following chemicals and apparatus:

- 0.100 mol dm^{-3} HCl solution;
- 1.0 g of impure BaCO$_3$;
- Gas syringe; and
- Standard glassware in the lab.

Q Where should we start thinking?

A: (i) What is the purpose of the plan?
— To determine the ideal gas constant, R, of the ideal gas equation, $pV = nRT$.

(ii) What do you need to know in order to determine the value of R?
— We need to know the number of moles of CO_2, the pressure, and the temperature.

(iii) How do you then determine the number of moles of CO_2?
— We can make use of the following balanced equation for the reaction between $BaCO_3$ and HCl:

$$BaCO_3(s) + 2HCl(aq) \rightarrow BaCl_2(aq) + CO_2(g) + H_2O(l).$$

Thus, if we know the number of moles of $BaCO_3$ that is used, we would be able to deduce the number of moles of CO_2 gas present from the volume of gas that we have collected.

(iv) How do you measure the temperature and pressure?
— Use a thermometer and a barometer, respectively.

(v) But how would you know how much of the sample to weigh?
— We can assume that we need to collect 50 cm^3 of the gas. From here, we can determine the number of moles of CO_2 present in the volume of the gas and then the number of moles of $BaCO_3$, hence calculating the mass of $BaCO_3$ that we need to use.

Pre-Experimental Calculations:

$$BaCO_3(s) + 2HCl(aq) \rightarrow BaCl_2(aq) + CO_2(g) + H_2O(l)$$

Assuming that 50 cm^3 of CO_2 gas is collected at r.t.p. conditions:

At r.t.p. (1 atm, 20°C), molar volume = 24.0 dm^3 mol^{-1}

Amount of CO_2 gas = $\frac{50}{24,000}$ = 2.08×10^{-3} mol

Amount of $BaCO_3$ present = Amount of CO_2 gas = 2.08×10^{-3} mol

Assuming that the sample consists only of $BaCO_3$:

Molar mass of $BaCO_3$ = 137.3 + 12.0 + 3(16.0) = 197.3 g mol^{-1}

Mass of $BaCO_3$ present = $2.08 \times 10^{-3} \times 197.3 = 0.4104$ g

Amount of HCl used = 2 × Amount of CO_2 gas = 4.16×10^{-3} mol

Volume of 0.100 mol dm^{-3} HCl solution needed = $\dfrac{4.16 \times 10^{-3}}{0.100}$ = 0.0416 dm^3
= 41.6 cm^3.

Do you see that the whole planning process here is exactly the same as that in Section 3.2? Thus, do take note of the pattern for such planning exercises. Do not be intimidated by the aim of the experiment. In essence, a lot of the thinking that is involved in planning experiments is very similar to one another.

Treatment of Results:

Let the volume of the CO_2 gas collected be V m^3.

Let the pressure of the gas collected be p Pa.

Let the temperature of the gas be T K.

Let the mass of $BaCO_3$ used be m g.

Molar mass of $BaCO_3$ = 137.3 + 12.0 + 3(16.0) = 197.3 g mol^{-1}

Amount of $BaCO_3$ used = $\dfrac{m}{197.3}$ mol

$BaCO_3(s) + 2HCl(aq) \rightarrow BaCl_2(aq) + CO_2(g) + H_2O(l)$

Amount of CO_2 gas = Amount of $BaCO_3$ used = $\dfrac{m}{197.3}$ mol

Hence, from pV = nRT $\Rightarrow$ R = $\dfrac{pV}{nT}$ = $\dfrac{197.3\, pV}{mT}$ J mol^{-1} K^{-1}.

Q So, the difference between Section 3.3 and Section 3.2 is that in Section 3.2, because the sample of $BaCO_3$ used is not pure, the number of moles of $BaCO_3$ used is an unknown. Hence, we need to *use the volume of gas that is collected to deduce the number of moles of $BaCO_3$ present through the balanced equation*. As for Section 3.3, the number of moles of $BaCO_3$ used is known. Hence, we can *make use of it to deduce the number of moles of CO_2 collected through a balanced equation*. Now, with the help of data on the pressure, temperature, and volume, we can determine the value of the gas constant, R. Am I right here?

A: Absolutely brilliant!

> Why is it not appropriate to use CO_2 to determine the ideal gas constant, R?

A: CO_2 is not an ideal gas because of its intermolecular forces. A better gas to use is probably H_2 gas which is slightly more ideal than CO_2 because of its weaker intermolecular forces.

3.4 Determine the Identity of an Acid

The Task:

You are given an acid of concentration, 9 g dm^{-3}. This acid can be one of the following monobasic acids:

- HA of M_r 42.9;
- HB of M_r 57.3; or
- HC of M_r 90.0.

Describe how you would determine the identity of the acid through the gas collection method. You are provided with the following chemicals and apparatus:

- 1 g of solid magnesium carbonate, $MgCO_3$;
- Gas syringe; and
- Standard glassware in the lab.

> Where should we start thinking?

A: (i) What is the purpose of the plan?
 — To determine the identity of the acid.
(ii) What do you need to know in order to determine the identity of the acid?
 — Since the concentration (9 g dm^{-3}) of the acid is known, we can obtain the molar concentration in mol dm^{-3}. This would mean that there are three possible concentrations for the acid depending on which acid it is:

- HA of concentration = $\dfrac{9}{42.9} = 0.210$ mol dm^{-3};
- HB of concentration = $\dfrac{9}{57.3} = 0.157$ mol dm^{-3}; or
- HC of concentration = $\dfrac{9}{90.0} = 0.100$ mol dm^{-3}.

Now, the acid would react with the MgCO$_3$ as follows:

$$2\text{HA/HB/HC} + \text{MgCO}_3(s) \rightarrow \text{MgA}_2/\text{MgB}_2/\text{MgC}_2 + \text{CO}_2(g) + \text{H}_2\text{O}(l).$$

If we know the number of moles of CO$_2$ gas that is collected, we would be able to calculate the number moles of the acid molecules that produced the CO$_2$ gas from the above balanced equation. With the volume of the acid known, we can determine its concentration and hence know the identity of the acid.

(iii) But, how would you know how much of the MgCO$_3$ to weigh?
— We can assume that we need to collect 50 cm^3 of the gas and that the acid is of the highest concentration among the three, i.e., 0.210 mol dm^{-3}.

Pre-Experimental Calculations:

$$2\text{HA/HB/HC} + \text{MgCO}_3(s) \rightarrow \text{MgA}_2/\text{MgB}_2/\text{MgC}_2 + \text{CO}_2(g) + \text{H}_2\text{O}(l)$$

Assuming that 50 cm^3 of CO$_2$ is collected at r.t.p. conditions:

At r.t.p. (1 atm, 20°C), molar volume = 24.0 dm^3 mol^{-1}

Amount of CO$_2$ gas = $\dfrac{50}{24{,}000} = 2.08 \times 10^{-3}$ mol

Amount of MgCO$_3$ present = Amount of CO$_2$ gas = 2.08×10^{-3} mol

Molar mass of MgCO$_3$ = 24.3 + 12.0 + 3(16.0) = 84.3 g mol^{-1}

Mass of MgCO$_3$ present = $2.08 \times 10^{-3} \times 84.3 = 0.1753$ g

Assuming that the concentration of the acid is due to HA, i.e., 0.210 mol dm^{-3}:

Amount of HA needed = 2 × Amount of CO$_2$ gas = 4.16×10^{-3} mol

Volume of HA solution needed = $\dfrac{4.16 \times 10^{-3}}{0.210} = 0.0198$ dm^3 = 19.8 cm^3.

Thus, if we use 0.1753 g of MgCO$_3$ and 19.8 cm^3 of the acid:

For HA, volume of CO_2 gas collected $= 50$ cm^3.

For HB: Amount of HB in 19.8 cm^3 = $\dfrac{19.8}{1000} \times 0.157 = 3.11 \times 10^{-3}$ mol

Amount of CO_2 collected $= \dfrac{1}{2} \times 3.11 \times 10^{-3} = 1.55 \times 10^{-3}$ mol

Volume of CO_2 collected $= 1.55 \times 10^{-3} \times 24{,}000 = 37.3$ cm^3.

For HC: Amount of HC in 19.8 cm^3 = $\dfrac{19.8}{1000} \times 0.100 = 1.98 \times 10^{-3}$ mol

Amount of CO_2 collected $= \dfrac{1}{2} \times 1.98 \times 10^{-3} = 9.90 \times 10^{-4}$ mol

Volume of CO_2 collected $= 9.90 \times 10^{-4} \times 24{,}000 = 23.8$ cm^3.

Hence, by measuring the volume of the gas collected, we can determine the identity of the acid. The procedure to carry out the experiment is similar to what we have discussed in the previous sections.

3.5 Determine the Value of n in $(COOH)_2 \cdot nH_2O$ Using Gas Collection

The Task:

Ethanedioic acid, $(COOH)_2 \cdot nH_2O$, undergoes a redox reaction with acidified $KMnO_4$:

$$2MnO_4^- + 5(COOH)_2 + 6H^+ \rightarrow 2Mn^{2+} + 10CO_2 + 8H_2O.$$

Describe how you would determine the value of n using the gas collection method. You are provided with the following chemicals and apparatus:

- 1 g of $(COOH)_2 \cdot nH_2O$;
- 0.010 mol dm^{-3} $KMnO_4$ solution;
- 0.05 mol dm^{-3} H_2SO_4 solution;
- Gas syringe; and
- Standard glassware in the lab.

Q Where should we start thinking?

A: (i) What is the purpose of the plan?
— To determine the value of n in $(COOH)_2 \cdot nH_2O$.

(ii) What do you need to know in order to determine the value of n in $(COOH)_2 \cdot nH_2O$?

— We need to know the number of moles of $(COOH)_2 \cdot nH_2O$ present in a fixed mass of the sample that we have measured as the amount in moles of $(COOH)_2 \cdot nH_2O$ is calculated as follows:

Amount of $(COOH)_2 \cdot nH_2O$ in moles

$$= \frac{\text{Mass of } (COOH)_2 \cdot nH_2O \text{ used}}{\text{Relative molecular mass of } (COOH)_2 \cdot nH_2O}$$

$$= \frac{\text{Mass of } (COOH)_2 \cdot nH_2O \text{ used}}{90 + 18n}$$

Hence, if both the amount of $(COOH)_2 \cdot nH_2O$ in moles and the mass of $(COOH)_2 \cdot nH_2O$ used are known, the value of n can be easily calculated.

(iii) How can you then determine the number of moles of $(COOH)_2 \cdot nH_2O$?

— We can make use of the following balanced equation for the reaction between $(COOH)_2 \cdot nH_2O$ and MnO_4^-:

$$2MnO_4^- + 5(COOH)_2 + 6H^+ \rightarrow 2Mn^{2+} + 10CO_2 + 8H_2O.$$

Thus, if we know the number of moles of CO_2 gas collected, we would be able to deduce the number of moles of $(COOH)_2$ present in the sample. Hence, the value of n in $(COOH)_2 \cdot nH_2O$ can be found as one mole of $(COOH)_2 \cdot nH_2O$ contains one mole of $(COOH)_2$ and n moles of H_2O.

Pre-Experimental Calculations:

$$2MnO_4^- + 5(COOH)_2 + 6H^+ \rightarrow 2Mn^{2+} + 10CO_2 + 8H_2O$$

Assuming that 50 cm^3 of CO_2 gas is collected at r.t.p. conditions:

At r.t.p. (1 atm, 20°C), molar volume = 24.0 dm^3 mol^{-1}

Amount of CO_2 gas = $\frac{50}{24,000}$ = 2.08 × 10^{-3} mol

Amount of $(COOH)_2$ present = $\frac{5}{10}$ × Amount of CO_2 gas = 1.04 × 10^{-3} mol

Molar mass of $(COOH)_2$ = 2(12.0 + 32.0 + 1.0) = 90.0 g mol^{-1}

Mass of $(COOH)_2$ present = 1.04 × 10^{-3} × 90 = 0.0936 g

Amount of MnO_4^- used = $\frac{2}{10}$ × Amount of CO_2 gas = 4.16 × 10^{-4} mol

Volume of MnO_4^- used = $\frac{4.16 \times 10^{-4}}{0.010}$ = 0.0416 dm^3 = 41.6 cm^3

Amount of H^+ needed = $\frac{6}{10}$ × Amount of CO_2 gas = 1.25×10^{-3} mol

Amount of H_2SO_4 needed = $\frac{1}{2} \times 1.25 \times 10^{-3} = 6.24 \times 10^{-4}$ mol

Volume of H_2SO_4 needed = $\frac{6.24 \times 10^{-4}}{0.05} = 1.25 \times 10^{-2}$ dm^3 = 12.5 cm^3.

Assuming that the value of n is zero, then the mass of $(COOH)_2$ is the "same" as the mass of $(COOH)_2 \cdot nH_2O$ used.

The Procedure:

Procedure for gas collection:

(i) Set up the apparatus as shown (refer to p. 77).
(ii) Accurately measure *0.0936 g* of the sample and place it in the conical flask.
(iii) Use a *measuring cylinder*, introduce *13.0 cm^3* of 0.05 mol dm^{-3} H_2SO_4 solution into the conical flask.
(iv) Record the *initial reading of the syringe*.
(v) Using the *graduated dropping funnel*, run in *42.0 cm^3* of 0.010 mol dm^{-3} $KMnO_4$ solution. Swirl to mix gently.
(vi) Allow the reaction to progress until the *plunger of the syringe does not move*.
(vii) Allow time for the volume of the gas to equilibrate to the temperature and pressure.
(viii) Record the *final reading of the syringe*. Remember to *subtract the volume of solution that is introduced from the final reading of the syringe*.
(ix) *Repeat the experiment* to check for *reliability of results*.

Note: In addition, such planning process can also be used for the following similar cases:

- Determine the concentration of H_2O_2 through the redox reaction using acidified $KMnO_4$:

$$2MnO_4^-(aq) + 6H^+(aq) + 5H_2O_2(aq)$$
$$\rightarrow 2Mn^{2+}(aq) + 5O_2(g) + 8H_2O(l)$$

We can collect the O_2 gas so as to determine the number of moles of H_2O_2.

3.6 Determine the Decomposition Equation of $RbNO_3$ Using Gas Collection

The Task:

Upon strong heating, rubidium nitrate, $RbNO_3$, decompose according to one of the following equations:

$$RbNO_3(s) \rightarrow \frac{1}{2}Rb_2O(s) + NO_2(g) + \frac{1}{4}O_2(g); \text{ or}$$

$$RbNO_3(s) \rightarrow RbNO_2(s) + \frac{1}{2}O_2(g).$$

Describe how you would determine the actual decomposition equation of $RbNO_3$ using the gas collection method. You are provided with the following chemicals and apparatus:

- 1 g of $RbNO_3$;
- Gas syringe; and
- Standard glassware in the lab.

Q Where should we start thinking?

A: (i) What is the purpose of the plan?
— To determine the actual decomposition equation of $RbNO_3$.

(ii) What do you need to know in order to determine the actual decomposition equation of $RbNO_3$?
— We need to know the number of moles of gaseous molecules collected for a fixed mass of $RbNO_3$ that has decomposed.

(iii) How can you then determine the number of moles of $RbNO_3$?
— When one mole of $RbNO_3$ is decomposed, with the following decomposition equation:

$$RbNO_3(s) \rightarrow \frac{1}{2}Rb_2O(s) + NO_2(g) + \frac{1}{4}O_2(g),$$

we would get 1.25 moles of gaseous molecules. But if the decomposition equation is as follows:

$$RbNO_3(s) \rightarrow RbNO_2(s) + \frac{1}{2}O_2(g),$$

we would simply get 0.5 moles of gaseous molecules. So, from the volume of the gas that is collected, we would be able to determine what is the actual decomposition equation.

(iv) But how do you know how much of the solid to use?

— Firstly, we need to assume that the decomposition equation is

$$RbNO_3(s) \rightarrow \frac{1}{2}Rb_2O(s) + NO_2(g) + \frac{1}{4}O_2(g).$$

By assuming this, it allows us to anticipate the maximum volume of the gas that we would collect.

Next, we need to assume that the volume of the gas that we are going to collect is 50 cm^3. By making this assumption, we would know the number of moles of gas particles and then deduce the amount of RbNO$_3$ that is needed to be decomposed.

Pre-Experimental Calculations:

Assuming that the decomposition equation is as follows:

$$RbNO_3(s) \rightarrow \frac{1}{2}Rb_2O(s) + NO_2(g) + \frac{1}{4}O_2(g).$$

Assuming that 50 cm^3 of gaseous molecules is collected at r.t.p. conditions:

At r.t.p. (1 atm, 20°C), molar volume = 24.0 dm^3 mol^{-1}

Amount of gaseous molecules = $\frac{50}{24,000}$ = 2.08 × 10^{-3} mol

Amount of NO$_2$ = $\frac{1}{1.25}$ × Amount of gaseous molecules = 1.66 × 10^{-3} mol

Amount of RbNO$_3$ present = Amount of NO$_2$ gas = 1.66 × 10^{-3} mol

Molar mass of RbNO$_3$ = 85.5 + 14.0 + 3(16.0) = 147.5 g mol^{-1}

Mass of RbNO$_3$ present = 1.66 × 10^{-3} × 147.5 = 0.2454 g.

Thus, if we use 0.2454 g of $RbNO_3$:

For $RbNO_3(s) \rightarrow \frac{1}{2}Rb_2O(s) + NO_2(g) + \frac{1}{4}O_2(g)$, volume of gas collected
$= 50 \text{ cm}^3$.

For $RbNO_3(s) \rightarrow RbNO_2(s) + \frac{1}{2}O_2(g)$:

Amount of $RbNO_3$ in 0.2454 g $= 1.66 \times 10^{-3}$ mol

Amount of O_2 collected $= \frac{1}{2} \times 1.66 \times 10^{-3} = 8.30 \times 10^{-4}$ mol

Volume of O_2 collected $= 8.30 \times 10^{-4} \times 24{,}000 = 19.9 \text{ cm}^3$.

Hence, by measuring the volume of the gas collected, we can determine the correct decomposition equation for $RbNO_3$.

The Procedure:

Procedure for gas collection:

(i) Set up the apparatus as shown below.
(ii) Accurately measure *0.2454 g of the sample in a boiling tube*.
(iii) Record the *initial reading of the syringe*.
(iv) Start *heating the solid gently then strongly* until the *plunger of the syringe does not move*.
(v) *Allow time for the volume of the gas to equilibrate to the temperature and pressure*.
(vi) Record the *final reading of the syringe*.
(vii) *Repeat the experiment* to check for *reliability of results*.

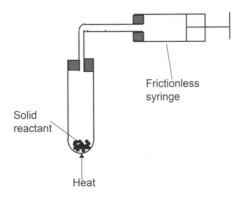

3.7 Determine the Concentration of H_2O_2 Using Gas Collection

The Task:

Hydrogen peroxide decomposes slowly in accordance to the following equation:

$$2H_2O_2(aq) \rightarrow 2H_2O(l) + O_2(g).$$

The reaction can be accelerated by using manganese(IV) oxide as the catalyst.

Given a sample of aqueous hydrogen peroxide in which the concentration is approximately in the range of 0.100 to 0.150 mol dm^{-3}, determine its actual concentration using the gas collection method. You are provided with the following chemicals and apparatus:

- 50 cm^3 of H_2O_2 solution;
- Solid manganese(IV) oxide;
- Gas syringe; and
- Standard glassware in the lab.

Q Where should we start thinking?

A: (i) What is the purpose of the plan?
— To determine the actual concentration of H_2O_2.

(ii) What do you need to know in order to determine the actual concentration of H_2O_2?
— We need to know the number of moles of O_2 molecules collected, which would give us the number of moles of H_2O_2 present in a specific volume of the solution used:

$$2H_2O_2(aq) \rightarrow 2H_2O(l) + O_2(g).$$

(iii) But how are you going to know how much of the H_2O_2 solution to use?
— Firstly, we need to assume that the concentration of the solution is 0.150 mol dm^{-3}. By assuming this, it allows us to anticipate the maximum volume of the gas that we would collect.

Next, we need to assume that the volume of the gas that is collected is 50 cm^3. By making this assumption, we would know the number of moles of gas particles that we have collected and then deduce the amount of H_2O_2 solution that is needed to be decomposed.

(iv) As the decomposition of H_2O_2 is rather slow, how can we speed it up?
— Well, we can add some manganese(IV) oxide as a catalyst.

Pre-Experimental Calculations:

$$2H_2O_2(aq) \rightarrow 2H_2O(l) + O_2(g)$$

Assuming that 50 cm^3 of gaseous molecules is collected at r.t.p. conditions:

At r.t.p. (1 atm, 20°C), molar volume = 24.0 dm^3 mol^{-1}

Amount of O_2 gas = $\frac{50}{24,000}$ = 2.08×10^{-3} mol

Amount of H_2O_2 = 2 × Amount of O_2 gas = 4.16×10^{-3} mol

Assuming that the concentration of the H_2O_2 solution is 0.150 mol dm^{-3}:

Volume of H_2O_2 solution needed = $\frac{4.16 \times 10^{-3}}{0.150}$ = 0.0277 dm^3 = 27.7 cm^3.

The Procedure:

Procedure for gas collection:

(i) Set up the apparatus as shown below:

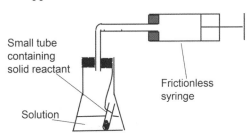

(ii) Use a 50 cm^3 *burette* to place *27.00 cm^3* of H_2O_2 solution into the conical flask.

(iii) Attached a *small tube containing some solid manganese(IV) oxide* as shown in the diagram. *Make sure that the catalyst does not come in contact with the solution.*

(iv) Record the *initial reading of the syringe*.
(v) *Let the manganese(IV) oxide come in contact with the solution by loosening the stopper. Swirl* to mix *gently*.
(vi) *Allow time for the volume of the gas to equilibrate to the temperature and pressure*.
(vii) Record the *final reading of the syringe*.
(viii) *Repeat the experiment* to check for *reliability of results*.

> **Q** We can actually determine the concentration of the H_2O_2 solution via titration; is it a better method than gas collection?

A: Certainly, the titration method is more accurate than the gas collection method as the accuracy of the latter greatly fluctuates with temperature and pressure. In addition, it is not easy to collect all the gas that has evolved. But nevertheless, the gas collection method is known for its ease of use and efficiency.

3.8 Safety Precautions for Gas Collection Experiments

You may be asked to quote some safety precautions while performing a gas collection experiment. Depending on the type of gas collection that you are performing, the following examples may be useful for you to take note:

— If the gas that you are collecting is an irritant or poisonous, such as NO_2, NH_3 or Cl_2 gases, conduct the experiment in a fumehood.
— If you are collecting the gas over water, there is a potential suck back of the water into the reaction vessel when heating is stopped. Thus, the delivery tube must be detached from the water bath when heating stops.

3.9 Minimizing Experimental Errors or Increasing Reliability

You may be asked to quote some ways to improve experimental reliability while performing a gas collection experiment. This method is fast but not very accurate. Thus, if there are more reliable methods instead of gas collection, the experimenter should avoid using the gas collection method. Nevertheless, depending on the type of gas collection that you are performing, the following examples may be useful for you to take note:

— The set-up must be properly clamped using retort stands.
— The set-up must be airtight to minimize the gas from escaping so that the result obtained is accurate.
— When the conical flask is swirled to mix the reactants, take note that the swirling would vary the volume of gas that is collected in the measuring cylinder/burette.
— As the gas is compressible, the heat of reaction may affect the volume of the gas collected. Thus, it is important to let the system equilibrate to the temperature and pressure first before reading the volume.
— Although the ideal gas equation may be used to determine the amount of gas particles collected, it is still not a good estimate as the gas is still non-ideal in nature.
— Add excess reagent to ensure that all the analytes have reacted so that the measurement is quantitative in nature.

CHAPTER 4

PLANNING FOR ENERGETICS EXPERIMENTS

During a chemical reaction, there may be an energy change. Such energy change affects (1) how fast the chemical reaction takes place, and (2) the extent of the chemical reaction, i.e., how complete the reaction will be. If the change of energy is in the form of heat flow, then through the measurement of the quantity of this heat, we can calculate the energy change for the reaction. But how can we measure this amount of heat change? Heat flows because there is a temperature gradient; heat is directly proportional to temperature. Hence, by measuring the temperature change, we can calculate the heat change.

But we need a medium to "shoulder" the heat that is evolved during the reaction — what is it that would "take in" the heat? Normally, water is the medium to shoulder the heat and with this, we can make use of the following equation to determine the heat change:

Heat change = $m \cdot c \cdot \Delta T$, where m = mass of the solution in g
c = heat capacity of the solution in J K^{-1} g^{-1}
ΔT = measured temperature change ($T_f - T_i$) in K or °C.

If water is the medium, and assuming that most of the heat energy goes to the water molecules which are present in a large amount, then the above equation is simplified to:

Heat change = $m \cdot c \cdot \Delta T$
= $\rho \cdot V \cdot c \cdot \Delta T$, where ρ = density of the solution in g cm^{-3}
and
V = volume of the solution in cm^3.

If we assume that the density of the solution is equivalent to that of pure water, i.e., 1 g cm^{-3}, then we would have:

Heat change = $V \cdot c \cdot \Delta T$, where c = heat capacity of the solution in J K^{-1} cm^{-3}.

When conducting a thermochemical experiment, the reactants are quickly mixed together as it is an "ideal" intention to let the reaction happen "at one go," where the heat energy would also be released "at one go." Hence, our thermometer would be able to register the temperature change in the shortest time with minimal heat lost to the surroundings. The experimental set-up used for this experiment can be one of the following:

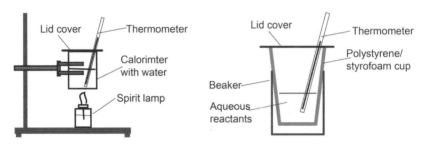

Q What else do we need to know when performing an energetics experiment?

A: It will be good if you understand some of the common enthalpy terms. Most importantly, you need to know how to construct an energy cycle. For more details, you can refer to *Understanding Advanced Physical Inorganic Chemistry* by J. Tan and K. S. Chan.

4.1 Determine the Enthalpy Change of Solution of Sodium Chloride

The Task:

Sodium chloride dissolves in water in accordance to the following equation:

$$NaCl(s) + 150H_2O(l) \rightarrow NaCl \cdot 150H_2O(aq).$$

You are supposed to plan an experiment to determine the enthalpy change for the above reaction with the following chemicals and apparatus:

- 5 g of solid NaCl;
- Deionized water;
- Styrofoam cup;
- Thermometer; and
- Standard glassware in the lab.

Q Where should we start thinking?

A: (i) What is the purpose of the plan?
— To determine the enthalpy change for the dissolving of NaCl in water.

(ii) What do you need to know in order to determine the enthalpy change?
— We need to know the temperature change for the reaction.

(iii) How do you then determine the temperature change?
— We can make use of the following balanced equation for the reaction between NaCl and H_2O:

$$NaCl(s) + 150H_2O(l) \rightarrow NaCl \cdot 150H_2O(aq).$$

Thus, if we know the number of moles of NaCl and H_2O used, we would be able to measure the temperature change.

(iv) How do you determine the number of moles of H_2O used?
— Well, if we assume that 50 cm^3 of water is used, we would be able to calculate its number of moles and then deduce the number of moles of NaCl needed. Thus, we would be able to calculate the mass of NaCl that we need to weigh.

Pre-Experimental Calculations:

$$NaCl(s) + 150H_2O(l) \rightarrow NaCl.150H_2O(aq)$$

Assuming that 50 cm^3 of H_2O is used:
If the density of water is 1 g cm^{-3}, the mass of 50 cm^3 of water is 50 g

Amount of H_2O used = $\frac{50}{18.0}$ = 2.778 mol

Amount of NaCl needed = $\frac{1}{150}$ × Amount of H_2O used = 1.85×10^{-2} mol

Molar mass of NaCl = $23.0 + 35.5 = 58.5$ g mol^{-1}

Mass of NaCl needed = $1.85 \times 10^{-2} \times 58.5 = 1.0833$ g.

The Procedure:

The general procedure for an energetics experiment involves the measurement of temperature change during a chemical reaction. Thus, when a student follows the experimental procedure diligently, he/she should be able to obtain repeatable results. Similarly, another student doing the same experiment should get data that are close to what the rest have obtained. Hence, the procedure for a typical energetics experiment consists of a series of steps which inform the student: (1) which step should come first; (2) what apparatus should he/she use; (3) what is the quantity of substance that he/she should measure; and (4) if needed, the reaction conditions such as temperature and pressure.

Procedure for measuring ΔT:

(i) Set up the apparatus as shown in the diagram (refer to p. 104).
(ii) Use a 50 cm^3 *burette* to introduce *50* cm^3 of deionized water into a *clean and dry styrofoam cup*.
(iii) Place the *thermometer* into the water and take note of the *initial temperature* of the water after some time.
(iv) *Weigh* accurately *1.0833* g of solid NaCl. Take note of the weight of the weighing bottle and the salt.
(v) *Empty the contents* of the weighing bottle into the styrofoam cup. *Stir gently* with the thermometer.
(vi) Take note of the *lowest drop* in temperature. (If the reaction is exothermic, then record the highest rise in temperature).
(vii) *Reweigh the emptied weighing bottle.*
(viii) *Repeat the experiment* to get *reliable results*.

Q Why must we wait for some time before taking the initial temperature of the water?

A: Well, the thermometer and water may not be of the same temperature. Thus, you need to wait for thermal equilibrium to set in.

Q Why must we stir gently?

A: We do not want you to stir too vigorously such that the contents are all spilled. In addition, the thermometer should not be scratching the wall of the cup as this would create frictional heat which would affect the accuracy of the experiment.

Q Why must we reweigh the emptied weighing bottle?

A: This is because you may not have quantitatively transferred everything from the weighing bottle into the cup. There would be some leftover solid in the weighing bottle.

Q Can we use some water to rinse the solid into the styrofoam cup?

A: No! Do not forget that the water in the cup has been accurately measured. Most importantly, when we perform a thermochemical experiment, after all the reactants are mixed, the priority is to get the temperature change in the shortest possible time so as to minimize heat gain or heat loss to the surroundings. Hence, continuous stirring is very important.

Results and Calculations:

Mass of weighing bottle and solid NaCl/g	
Mass of **emptied** weighing bottle/g	
Mass of solid NaCl used/g	x

Lowest temperature reached/°C	T_f
Initial temperature/°C	T_i
Decrease in temperature (ΔT)/°C	$T_i - T_f$

Assume that the specific heat capacity of water is 4.18 J g^{-1} K^{-1}.
Assume that no heat is absorbed from surroundings.

Heat change = $m \cdot c \cdot \Delta T$
  = 50.0 × 4.18 × $(T_i - T_f)$ = 209$(T_i - T_f)$ J   {positive for heat change since it is an endothermic reaction}

Amount of NaCl = $\frac{x}{58.5}$ mol

Hence, enthalpy change = $\frac{\text{Heat change}}{\text{Amount of NaCl}}$

  = $\{209(T_i - T_f)\} \div \frac{x}{58.5}$ J mol^{-1}.

Q So, we do not include the mass of the solid NaCl used when we calculate the value of heat change using 'heat change = $m \cdot c \cdot \Delta T$?'

A: You don't have to do that. As mentioned before, the heat absorbed (for this case) during the reaction comes mostly from the water molecules as these are present in large quantities as compared to the sodium chloride ions. At this point, many students will make the mistake by adding the mass of the solid to the water when they use the equation 'heat change = $m \cdot c \cdot \Delta T$.' The students would think that the mass, m, includes the mass of the solid that has been added.

Q What other solids can also make use of the same planning process outlined above?

A: Many! For example, you can determine the enthalpy change of solution of $NaHCO_3$, K_2CO_3, NH_4NO_3, etc. The main idea is to find out how much water is to be used, then how much solid is to be dissolved and thereafter, measure the temperature change during the reaction in the shortest time possible.

4.2 Determine the Enthalpy Change of Reaction Between Aqueous Na_2CO_3 and Aqueous HCl

The Task:

Aqueous sodium carbonate reacts with aqueous hydrochloric acid as follows:

$$Na_2CO_3(aq) + 2HCl(aq) \rightarrow 2NaCl(aq) + H_2O(l) + CO_2(g).$$

You are supposed to plan an experiment to determine the enthalpy change for the above reaction with the following chemicals and apparatus:

- 1.00 mol dm^{-3} HCl solution;
- 0.50 mol dm^{-3} Na_2CO_3 solution;
- Styrofoam cup;
- Thermometer; and
- Standard glassware in the lab.

Q Where should we start thinking?

A: (i) What is the purpose of the plan?
— To determine the enthalpy change for the reaction between $Na_2CO_3(aq)$ and HCl(aq).

(ii) What do you need to know in order to determine the enthalpy change?
— We need to know the temperature change for the reaction.

(iii) How do you then determine the temperature change?
— We can make use of the following balanced equation for the reaction between $Na_2CO_3(aq)$ and HCl(aq):

$$Na_2CO_3(aq) + 2HCl(aq) \rightarrow 2NaCl(aq) + H_2O(l) + CO_2(g).$$

Thus, if we know the number of moles of Na_2CO_3 and HCl used, we would be able to measure the temperature change for the reaction.

(iv) How do you determine the number of moles of Na_2CO_3 and HCl used?
— Well, if we assume that 30 cm^3 of Na_2CO_3 solution is used, we would be able to calculate the number of moles of Na_2CO_3 that is present and then deduce the number of moles of HCl needed.

(v) So, is the ratio of Na_2CO_3 and HCl going to be exactly 1:2?
— No! In order to ensure that the reaction is completed, one of the reagents must be limiting. In this case, we let the HCl used be in excess.

Pre-Experimental Calculations:

$$Na_2CO_3(aq) + 2HCl(aq) \rightarrow 2NaCl(aq) + H_2O(l) + CO_2(g)$$

Assuming that 30 cm^3 of Na_2CO_3(aq) solution is used:

Amount of Na_2CO_3 used = $\frac{30}{1000} \times 0.50 = 0.015$ mol

Amount of HCl needed = $2 \times$ Amount of Na_2CO_3 used = 0.030 mol

Volume of HCl needed = $\frac{0.030}{1.00} = 0.030$ dm^3 = 30.0 cm^3.

The Procedure:

Procedure for measuring ΔT:

(i) Set up the apparatus as shown in the diagram (refer to p. 104).
(ii) Use a 50 cm^3 *burette* to introduce *30 cm^3* of Na_2CO_3 solution into a *clean and dry styrofoam cup*.
(iii) Place the *thermometer* into the solution and take note of the *initial temperature* of the solution after some time.
(iv) Use a *measuring cylinder*, introduce *32 cm^3* of HCl solution into the styrofoam cup.
(v) *Stir gently* with the thermometer.
(vi) Take note of the *highest rise in temperature*.
(vii) *Repeat the experiment* to get *reliable results*.

> **Q** Why do you use a measuring cylinder to introduce the HCl solution? Can we use a pipette or burette instead?

A: As the solution is in excess, there is no point in using a more accurate apparatus to measure the volume. Thus, we used a measuring cylinder. In addition, do you remember that in a thermochemical experiment, we need the reaction to occur at the shortest time so that the heat change happens in the shortest time? The measuring cylinder is thus better than the pipette or burette as it allows the solution to be emptied into the cup in the shortest time. So, remember, the last solution to add into the styrofoam cup, which would trigger the reaction, must be measured using a measuring cylinder.

Q When we calculate the heat change using 'heat change = $m \cdot c \cdot \Delta T$,' what is the value of the mass, m?

A: Since you have added two solutions together, the value of m is the total volume of the two solutions added = 30 + 32 = 62 g. Take note that this calculation of m is on the basis that the volume of the solution is 62 cm^3 and the density of the solution is assumed to be 1 g cm^{-3}.

Q So, when we calculate the enthalpy change for the reaction, the number of moles of the reactant used in the denominator of 'enthalpy change = $\dfrac{\text{Heat change}}{\text{Amount of limiting reagent}}$, is the amount of limiting reagent and not the amount of excess reagent?

A: Absolutely spot on!

Q We are adding another solution into the cup which has already contained another solution with the initial temperature being known. What happens when the initial temperature of the second solution is not the same as the one that is already inside the cup? Wouldn't there be thermal dilution?

A: Good thinking! Yes, there would be thermal dilution. So, what you can do is to measure the initial temperatures of both solutions before mixing. If they are the same, great! If not, simply apply the following equation to determine a weighted average of the initial temperature for the mixture:

Average initial temperature after mixing solution 1 and solution 2
$$= \dfrac{V_1 \times T_1 + V_2 \times T_2}{V_1 + V_2}.$$

> **Q:** Can we use the above plan to determine the enthalpy change for the following reaction?
>
> $$Na_2CO_3(s) + 2HCl(aq) \rightarrow 2NaCl(aq) + H_2O(l) + CO_2(g)$$

A: Theoretically yes, but practically, the experiment is highly inaccurate. Why? This is because the reaction of a solid carbonate and an acid would result in effervescence, causing acid spray and loss of the solid (refer to Chapter 1, Section 1.3). This situation of acid spray is going to be more serious than if you have mixed two aqueous solutions together. Thus, the measured result would not be useful.

> **Q:** So, is there a way to calculate the enthalpy change for the following reaction?
>
> $$Na_2CO_3(s) + 2HCl(aq) \rightarrow 2NaCl(aq) + H_2O(l) + CO_2(g)$$

A: You can make good use of the following energy cycle:

$$
\begin{array}{ccc}
Na_2CO_3(s) & \xrightarrow{\Delta H_{solution}} & Na_2CO_3(aq) \\
\Delta H_2 \downarrow + 2HCl(aq) & & \Delta H_1 \downarrow + 2HCl(aq) \\
& 2NaCl(aq) + H_2O(l) + CO_2(g) &
\end{array}
$$

Thus, if you know the enthalpy change of solution of solid Na_2CO_3, which can be easily determined using the plan in Section 4.1, you can make use of Hess's Law to calculate the enthalpy change for $Na_2CO_3(s) + 2HCl(aq) \rightarrow 2NaCl(aq) + H_2O(l) + CO_2(g)$:

By Hess's Law, $\Delta H_2 = \Delta H_{solution} + \Delta H_1$.

> **Q:** Wow! What is Hess' Law?

A: Hess's Law is actually an expression of the more general Law of Conservation of Energy, which states that energy cannot be created nor destroyed but transferred. So, Hess's Law states that the enthalpy change for a chemical process is only dependent on the initial and final states but independent of the pathway taken.

Note: You can also use the same planning outline for:
 (i) Determine the standard enthalpy change of neutralization.
    — *Standard enthalpy change of neutralization* is the energy change when a certain amount of acid neutralizes a base to *form 1 mole of water* (in dilute aqueous solution) at 298 K and 1 bar:
    Examples:
    $HCl(aq) + NaOH(aq) \rightarrow NaCl(aq) + H_2O(l)$
        {between a strong acid and a strong base}
    $H_2SO_4(aq) + 2NaOH(aq) \rightarrow Na_2SO_4(aq) + 2H_2O(l)$
        {between a dibasic strong acid and a strong base}
    $HCl(aq) + Ba(OH)_2(aq) \rightarrow BaCl_2(aq) + 2H_2O(l)$
        {between a diacidic strong base and a strong acid}
    $CH_3COOH(aq) + NaOH(aq) \rightarrow NaCH_3COO(aq) + H_2O(l)$
        {between a weak acid and a strong base}
    $CH_3COOH(aq) + NH_3(aq) \rightarrow NH_4CH_3COO(aq)$
        {between a weak acid and a weak base}
    $NH_3(aq) + HCl(aq) \rightarrow NH_4Cl(aq)$
        {between a weak base and a strong acid}

    The formula to use is 'enthalpy change = $\dfrac{\text{Heat change}}{\text{Amount of water formed in moles}}$'.
    Thus, although the reactants may not be the same, the methods are very similar. The biggest difference lies in the magnitude of the enthalpy change of neutralization that is obtained. For weak acid–strong base or weak base–strong acid or weak base–weak acid neutralization, the enthalpy change is less exothermic than for a strong acid–strong base because the energy that is given off during neutralization has been partly diverted to help dissociate the undissociated weak acid or weak base. For more details, you can refer to *Understanding Advanced Physical Inorganic Chemistry* by J. Tan and K. S. Chan.
 (ii) Determine the enthalpy change of the following reaction:
    $$CO_2(g) + KOH(aq) \rightarrow KHCO_3(aq).$$

The Task:

Plan two experiments to determine the enthalpy changes of the following reaction:
$$KHCO_3(aq) + HCl(aq) \rightarrow KCl(aq) + H_2O(l) + CO_2(g);$$
$$KOH(aq) + HCl(aq) \rightarrow KCl(aq) + H_2O(l).$$

> Then apply Hess's Law to calculate the enthalpy change of the following reaction:
>
> $$CO_2(g) + KOH(aq) \rightarrow KHCO_3(aq).$$

Q Instead of using HCl for HCl(aq) + Ba(OH)$_2$(aq) → BaCl$_2$(aq) + 2H$_2$O(l), can we replace the HCl solution with aqueous H$_2$SO$_4$ since both are strong acids?

A: No! The precipitation of the insoluble BaSO$_4$ would have a particular enthalpy change by itself. Hence, this may influence the actual enthalpy change that we are trying to measure. In essence, when we measure enthalpy change, there can only be one specific reaction taking place, and not more than one.

4.3 Determine the Enthalpy Change of Reaction Between Solid MgO and Aqueous HCl

The Task:

Solid magnesium oxide reacts with aqueous hydrochloric acid as follows:

$$MgO(s) + 2HCl(aq) \rightarrow MgCl_2(aq) + H_2O(l).$$

You are supposed to plan an experiment to determine the enthalpy change for the above reaction with the following chemicals and apparatus:

- 3 g of solid MgO;
- 0.50 mol dm^{-3} HCl solution;
- Styrofoam cup;
- Thermometer; and
- Standard glassware in the lab.

Q Where should we start thinking?

A: (i) What is the purpose of the plan?
— To determine the enthalpy change for the reaction between MgO(s) and HCl(aq).

(ii) What do you need to know in order to determine the enthalpy change?
— We need to know the temperature change for the reaction.
(iii) How do you then determine the temperature change?
— We can make use of the following balanced equation for the reaction between MgO(s) and HCl(aq):
$$MgO(s) + 2HCl(aq) \rightarrow MgCl_2(aq) + H_2O(l).$$
Thus, if we know the number of moles of MgO and HCl used, we would be able to measure the temperature change.
(iv) How do you determine the number of moles of MgO and HCl used?
— Well, if we assume that 50 cm^3 of HCl solution is used, we would be able to calculate the number of moles of HCl present and then deduce the number of moles of MgO that is needed.
(v) So, is the ratio of MgO and HCl going to be exactly 1:2?
— No! In order to ensure that the reaction is completed, one of the reagents must be limiting. In this case here, we let the HCl used be in excess.

Q Have you noticed from Sections 4.2 and 4.3, if one of the reactant is a solid, we would start the pre-calculation by assuming that a particular volume of solution is used instead of assuming that a particular amount of solid is used?

A: Yes. In Section 4.2, we assumed that 50 cm^3 of water is used. Here in Section 4.3, we assumed that we are going to use 50 cm^3 of the acid.

Q But why is it preferred to assume the volume of solution used in the pre-calculation rather than the mass of the solid?

A: If we assume that a particular volume of solution is used, we must make sure this volume is sufficient to submerge the bulb of thermometer. Thus, 50 cm^3 is a good consideration. But if we start by assuming a particular mass of solid is used, after the calculation, the volume of the solution may be too small to submerge the bulb of the thermometer.

Pre-Experimental Calculations:

$$MgO(s) + 2HCl(aq) \rightarrow MgCl_2(aq) + H_2O(l)$$

Assuming that 50 cm^3 of HCl(aq) solution is used:

Amount of HCl used = $\frac{50}{1000} \times 0.50 = 0.025$ mol

Amount of MgO needed = $\frac{1}{2} \times$ Amount of HCl used = 0.0125 mol

Molar mass of MgO = 24.3 + 16.0 = 40.3 g mol^{-1}

Mass of MgO needed = 0.0125 × 40.3 = 0.5038 g.

The Procedure:

Procedure for measuring ΔT:
  (i) Set up the apparatus as shown in the diagram (refer to p. 104).
  (ii) Use a *50 cm^3 burette* to introduce *60 cm^3* of HCl solution into a *clean and dry styrofoam cup*.
  (iii) Place the *thermometer* into the solution and take note of the *initial temperature* of the solution after some time.
  (iv) *Weigh* accurately *0.5038 g* of solid *MgO*. Take note of the weight of the weighing bottle and the salt.
  (v) *Empty* the contents of the weighing bottle into the styrofoam cup. *Stir gently* with the thermometer.
  (vi) Take note of the *highest rise* in temperature. (If the reaction is endothermic, then record the lowest drop in temperature).
  (vii) *Reweigh* the *emptied weighing bottle*.
  (viii) *Repeat the experiment* to get *reliable results*.

> **Q** Why is it preferred to have the aqueous HCl solution in excess instead of the solid MgO?

A: Good question! If the solid MgO is in excess, after all the HCl has reacted, the leftover solid MgO may react with the water molecules in the aqueous medium. Although MgO is not very soluble in water, the risk that this reaction would occur still exists. Hence, this would introduce unnecessary energy changes to our intended experiment.

Q Can we use a measuring cylinder to measure the volume of the aqueous HCl solution instead?

A: Well, in this case, using a burette to measure the volume of the solution would improve the accuracy. As both the volume and mass of the reactants would be accurately measured, why not use the burette?

Note: In addition, such a planning process can also be used for the following similar cases involving a solid and a solution:

(i) Determine the enthalpy change for the reaction of Mg with aqueous HCl (Or any other metal with any other acid as there are too many possible combinations):

$$Mg(s) + 2HCl(aq) \rightarrow MgCl_2(aq) + H_2(g);$$
$$Mg(s) + H_2SO_4(aq) \rightarrow MgSO_4(aq) + H_2(g).$$

Importantly, the reaction between the metal and acid cannot be too vigorous resulting in too much heat loss to the surroundings or reactant loss. For example, the reaction of sodium metal with acid is not encouraged.

In addition, if a metal strip is used instead of the powdered form, the result may vary because the smaller surface area of the metal strip would give us a slower rate of reaction. Hence, the longer reaction time allows more heat to be lost to the surroundings.

(ii) Determine the enthalpy change for the displacement reaction between solid zinc and copper(II) sulfate solution:

$$Zn(s) + Cu^{2+}(aq) \rightarrow Cu(s) + Zn^{2+}(aq).$$

(iii) Determine the enthalpy change of reaction between a Group 2 hydroxide with HCl solution:

$$M(OH)_2(s) + 2HCl(aq) \rightarrow MCl_2(aq) + 2H_2O(l), \quad M = Mg, Ca, Sr, Ba.$$

Plot a graph of the enthalpy change of solution of the Group 2 hydroxides with the respective hydroxides on the x-axis.

Q In all the previous thermochemical experiments that we have discussed, the experiments are performed in a styrofoam cup which cannot completely prevent any heat loss or heat gain. Is there a way to perform a thermochemical experiment that would take into consideration the amount of heat gain or heat loss?

A: Yes, of course there is. Refer to the following section.

4.4 Determine the Enthalpy Change of Reaction Between Solid MgO and Aqueous HCl by Plotting a Graph

The Task:

Solid magnesium oxide reacts with aqueous hydrochloric acid as follows:

$$MgO(s) + 2HCl(aq) \rightarrow MgCl_2(aq) + H_2O(l).$$

Using a graphical method, plan an experiment to determine the enthalpy change for the above reaction with the following chemicals and apparatus:

- 3 g of solid MgO;
- 0.50 mol dm^{-3} HCl solution;
- Styrofoam cup;
- Thermometer;
- Stopwatch; and
- Standard glassware in the lab.

Q If you compare the apparatus list here and the one in Section 4.3, what do you notice?

A: There is an additional stopwatch.

So, take note that all the preplanning would be the same! The main difference only lies in the procedure. So, all experiments that we have discussed before or we will be discussing can simply convert its original experimental procedures to the one as shown below:

Procedure for measuring ΔT:

(i) Set up the apparatus as shown in the diagram (refer to p. 104).
(ii) Use a *50 cm^3 burette* to introduce *60 cm^3* of HCl solution into a *clean and dry styrofoam cup*.
(iii) Place the *thermometer* into the solution and *start the stopwatch*.

(iv) *At regular time intervals* of 0.5 min, record the *temperature of the solution*.
(v) *Weigh* accurately *0.5038* g of solid MgO. Take note of the weight of the weighing bottle and the salt.
(vi) *Empty* the contents of the weighing bottle into the styrofoam cup at the *3*-min *point. Stir* gently with the thermometer.
(vii) *Do not record the temperature at the 3*-min *point*. But *record the temperature at 0.5* min *intervals after the 3*-min *point for about 10 min with continuous stirring*.
(viii) *Repeat the experiment* to get *reliable results*.

> **Q** How does the recording of the temperatures at regular time intervals help to correct for heat loss or heat gain?

A: When you plot the temperature versus time graph for an exothermic experiment, you get a profile which looks like this:

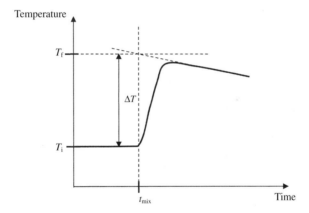

When you perform an extrapolation and obtain the maximum temperature at the point of mixing, which is at the 3-min point, the difference between the initial temperature and the maximum temperature is the theoretical change of temperature without any heat loss to the surroundings. This means that the vertical "jump" at the 3-min point is the point where the reaction has ideally happened and completed at "one go."

For an endothermic reaction, the process of extrapolation would look like the following:

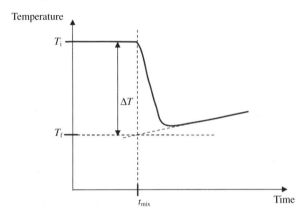

> **Q** What causes the subsequent increase in temperature after the minimum temperature?

A: This is due to the absorption of heat energy by the solution from the surroundings.

> **Q** Does this mean that if the styrofoam cup is "extremely" well-insulated, there would not be any increase in temperature?

A: Theoretically, yes. Similarly, if the cup is very "extremely" well-insulated, there would not be any heat loss for an exothermic reaction after the mixture has reached its maximum temperature rise as shown below:

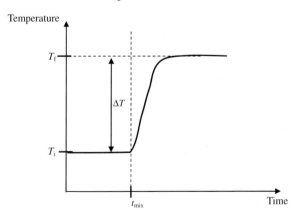

For an endothermic reaction, if the system is "extremely" well-insulated, the temperature versus time plot will be as follows:

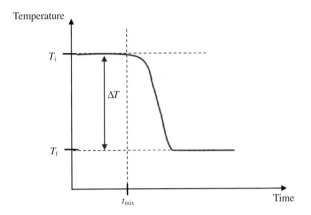

> **Q** What is the disadvantage of monitoring the temperature with respect to time compared to simply measuring the maximum rise in temperature?

A: Although monitoring temperature with respect to time allows us to correct for the heat lost to the surroundings or heat gained from the surroundings, this method is much more time consuming as compared to if we simply measure the maximum rise in temperature.

4.5 Determine the Enthalpy Change of Formation of MgO

The Task:

Solid magnesium oxide reacts with aqueous hydrochloric acid as follows:

$$MgO(s) + 2HCl(aq) \rightarrow MgCl_2(aq) + H_2O(l);$$
$$Mg(s) + 2HCl(aq) \rightarrow MgCl_2(aq) + H_2(g).$$

Given the enthalpy change of formation of water, $\Delta H_f(H_2O)$, plan an experiment to determine the enthalpy change of formation of MgO from Mg and O_2 with the following chemicals and apparatus:

- 3 g of solid MgO;
- 3 g of magnesium metal;

- 0.50 mol dm^{-3} HCl solution;
- Styrofoam cup;
- Thermometer; and
- Standard glassware in the lab.

After obtaining the enthalpies change for the following reactions using the plan in Section 4.3:

$$MgO(s) + 2HCl(aq) \rightarrow MgCl_2(aq) + H_2O(l);$$
$$Mg(s) + 2HCl(aq) \rightarrow MgCl_2(aq) + H_2(g),$$

you can make use of the results to calculate the enthalpy change of formation of MgO, which is also known as the enthalpy change of combustion of Mg, through an energy cycle:

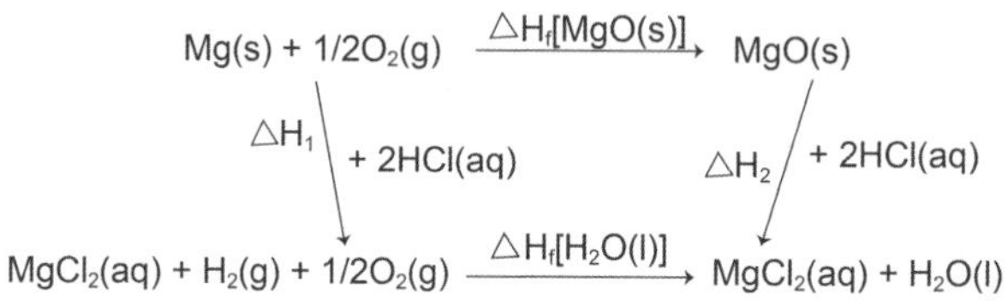

```
                   ΔH_f[MgO(s)]
Mg(s) + 1/2O_2(g) ─────────────→ MgO(s)
      │                              │
  ΔH_1│ + 2HCl(aq)          ΔH_2 │ + 2HCl(aq)
      ↓                              ↓
                   ΔH_f[H_2O(l)]
MgCl_2(aq) + H_2(g) + 1/2O_2(g) ──────→ MgCl_2(aq) + H_2O(l)
```

By Hess's Law, $\Delta H_f[MgO(s)] = \Delta H_1 + \Delta H_f[H_2O(l)] - \Delta H_2$.

4.6 Determine the Identity of an Acid using Thermometry

The Task:

The enthalpy change of neutralization between a strong monobasic acid and a strong monoacidic base to produce one mole of water is given as follows:

$$H^+(aq) + OH^-(aq) \rightarrow H_2O(l), \quad \Delta H = -57.0 \text{ kJ mol}^{-1}.$$

You are given a strong acid of concentration, 49 g dm^{-3}. This acid can be one of the following monobasic acids:

- HA of M_r 42.9;
- HB of M_r 57.3; or
- HC of M_r 90.0.

Describe how you would determine the identity of the acid through thermochemical means. You are provided with the following chemicals and apparatus:

- Acid solution of concentration, 49 g dm^{-3}, which can either be HA, HB or HC;
- 1.144 mol dm^{-3} of sodium hydroxide solution;
- Styrofoam cup;
- Thermometer; and
- Standard glassware in the lab.

Q Where should we start thinking?

A: (i) What is the purpose of the plan?
— To determine the identity of the acid.

(ii) What do you need to know in order to determine the identity of the acid?
— As each possible acid has its own concentration value, if we let a particular amount of NaOH react with a particular volume of the acid, there would be a particular amount of heat energy evolved as different numbers of moles of water would form. This amount of heat energy evolved would give us a particular ΔT value. If we now compare this ΔT value with the expected ΔT value for each of the possible acids, we would be able to determine the actual identity of the acid.

Pre-Experimental Calculations:

Since the concentration (49 g dm^{-3}) of the acid is known, we can obtain the molar concentration in mol dm^{-3}. This would mean that there are three possible molar concentrations depending on which acid it is:

- HA of concentration = $\frac{49}{42.9}$ = 1.142 mol dm^{-3};
- HB of concentration = $\frac{49}{57.3}$ = 0.855 mol dm^{-3}; or
- HC of concentration = $\frac{49}{90.0}$ = 0.544 mol dm^{-3}.

If we each take 25.0 cm³ of the acid:
- Amount of HA present = $\frac{25}{1000} \times 1.142 = 2.86 \times 10^{-2}$ mol;
- Amount of HB present = $\frac{25}{1000} \times 0.855 = 2.14 \times 10^{-2}$ mol; or
- Amount of HC present = $\frac{25}{1000} \times 0.544 = 1.36 \times 10^{-2}$ mol.

Thus, if we add 2.86×10^{-2} mol of NaOH from the 1.144 mol dm^{-3} NaOH solution into 25.0 cm³ the unknown acid solution:

$$H^+(aq) + OH^-(aq) \rightarrow H_2O(l).$$

Volume of NaOH solution = $\frac{2.86 \times 10^{-2}}{1.144} = 0.025$ dm³ = 25 cm³

Total volume of the solution = 25.0 + 25.0 = 50.0 cm³

Heat capacity of the solution is 4.18 J g^{-1} K^{-1} or 4.18 J cm^{-3} K^{-1}

For HA, the amount of H$_2$O formed = 5.25×10^{-3} mol

Enthalpy change, $\Delta H = \frac{\text{Heat evolved}}{\text{Amount of water formed in moles}} = -57.0$ kJ mol^{-1}

Heat evolved = $57.0 \times 2.86 \times 10^{-2} = 1.630$ kJ = 1630 J

$\Delta T = \frac{\text{Heat evolved}}{V \times c} = \frac{1630}{50 \times 4.18} = 7.8°C.$

For HB, the amount of H$_2$O formed = 2.14×10^{-2} mol
Heat evolved = $57.0 \times 2.14 \times 10^{-2} = 1.220$ kJ = 1220 J

$\Delta T = \frac{\text{Heat evolved}}{V \times c} = \frac{1220}{50 \times 4.18} = 5.8°C.$

For HC, the amount of H$_2$O formed = 1.36×10^{-2} mol
Heat evolved = $57.0 \times 1.36 \times 10^{-2} = 0.775$ kJ = 775 J

$\Delta T = \frac{\text{Heat evolved}}{V \times c} = \frac{775}{50 \times 4.18} = 3.7°C.$

Hence, by measuring the ΔT for the reaction, we can determine the identity of the acid!

The Procedure:

Procedure for measuring ΔT:

(i) Set up the apparatus as shown in the diagram (refer to p. 104).
(ii) Use a *pipette* to introduce *25.0 cm³* of the acid solution into a *clean and dry styrofoam cup*.

(iii) Place the *thermometer* into the solution and take note of the *initial temperature* of the solution after some time.
(iv) Use a *measuring cylinder* to introduce 25 cm³ of 1.144 mol dm⁻³ of sodium hydroxide solution into the styrofoam cup.
(v) *Stir gently* with the thermometer.
(vi) Take note of the *highest rise* in temperature.
(vii) *Calculate the ΔT* and see whether the value is close to 7.8°C (HA) or 5.8°C (HB) or 3.7°C (HC).
(viii) *Repeat the experiment* to get *reliable results*.

 Can we use a measuring cylinder to measure the volume of the acid solution instead?

A: Well, you have a choice to measure the volume of one of the two solutions accurately, so why not do it *with a pipette or a burette*? This would improve the accuracy of the result.

Note: In addition, such a planning process can also be used for the following similar cases:

(i) Determine the identity of a base using thermometry:

The Task:

The enthalpy change of neutralization between a strong monobasic acid and strong monoacidic base to produce one mole of water is given as follows:

$$H^+(aq) + OH^-(aq) \rightarrow H_2O(l) \qquad \Delta H = -57.0 \text{ kJ mol}^{-1}.$$

You are given a strong monoacidic base of concentration, 49 g dm⁻³. This base can be one of the following monoacidic bases:

- AOH of M_r 42.9;
- BOH of M_r 57.3; or
- COH of M_r 90.0.

(ii) Determine the value of n of the acid, H_nX, using thermometry/ Determine the basicity of the acid, H_nX, using thermometry:

The Task:

You are given a strong acid, H_nX (where $n = 1$ or 2), of concentration 0.855 mol dm^{-3}. Describe how you would determine the value of n using thermometry.

(The thinking behind is very similar to what we have discussed in Section 4.6. If $n = 1$, i.e., HX, then concentration of H^+ is 0.855 mol dm^{-3}. But if $n = 2$, i.e., H_2X, then concentration of H^+ is 1.71 mol dm^{-3}. So, it is similar to Section 4.6 as the given acid (H_nX) solution should be treated as having two possible concentrations.)

4.7 Determine the Concentration of an Acid Through Thermometric Titration

The Task:

A student was given a monobasic strong acid solution (HX) of concentration approximately 1 mol dm^{-3}. The reaction of the strong acid with sodium hydroxide solution is exothermic in nature. You are supposed to plan an experiment using thermometric titration, i.e., measuring the temperature of the mixture while adding the titrant, to determine the actual concentration of the acid with the following chemicals and apparatus:

- HX solution of concentration approximately 1 mol dm^{-3};
- 1.00 mol dm^{-3} NaOH solution;
- Styrofoam cup;
- Thermometer; and
- Standard glassware for titration.

Q Where should we start thinking?

A: (i) What is the purpose of the plan?
— To determine the concentration of the acid, HX.

(ii) What do you need to know in order to determine the concentration of the acid?
— We need to know the number of moles of HX present in a fixed volume of the sample solution that we have used.

(iii) How would you then determine the number of moles of HX?
— We need to know the end-point of the titration.

(iv) How do you get the end-point of the thermometric titration?
— Since the reaction between the acid and NaOH is exothermic, as we add more NaOH solution to a fixed volume of the acid, more heat energy is given off. Thus, the temperature of the solution is going to increase. At the end-point, a stoichiometric amount of NaOH would have been added, which means that the temperature rise is going to be at the maximum. After the end-point, if we continue to add NaOH solution, the temperature of the solution mixture is going to decrease because of thermal dilution. Hence, from the intersection point, we can determine the number of moles of NaOH needed and deduce the number of moles of HX present in accordance to the following equation:

$$HX(aq) + NaOH(aq) \rightarrow NaX(aq) + H_2O(l),$$
$$\Delta H = \text{negative value}.$$

The graphical plot for both an exothermic and endothermic reaction are shown below:

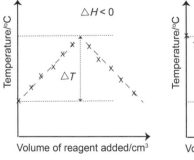
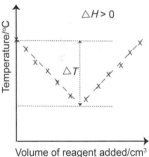

Q: Is there thermal dilution before the intersection point for an exothermic reaction when the titrant is added to the analyte?

A: Yes, certainly. But the thermal dilution is not significant enough to cause the temperature to drop as the reaction has not completed yet.

Q So, the increase in temperature after the intersection point for an endothermic reaction is due to the higher temperature of the titrant being added than the solution in the styrofoam cup?

A: You are right. Well done!

The Procedure:

Procedure for measuring temperature rise:

(i) Set up the apparatus as shown in the diagram below.
(ii) Use a *pipette* to introduce *25.0* cm^3 of the acid solution into a *clean and dry styrofoam cup*.
(iii) Place the *thermometer* into the solution and take note of the *initial temperature* of the solution after some time.
(iv) Use a 50 cm^3 *burette* to introduce *5.00* cm^3 *of 1.00* mol dm^{-3} *NaOH* solution.
(v) *Stir gently* with the thermometer. Take note of the *highest rise* in temperature.
(vi) *Repeat steps (iv) to (v)* until a total volume of *40* cm^3 of NaOH solution has been added.
(vii) *Plot a graph* of temperature rise versus volume of NaOH solution added. The *intersection point* will give the volume of NaOH solution that is needed to *completely react* with the acid.
(viii) *Repeat the experiment* to get *reliable results*.

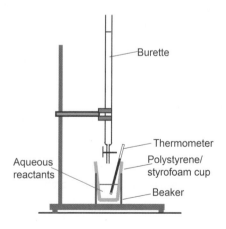

Q If we use two times the volume of each of the two solutions for the reaction, would the maximum temperature rise also be two times more than that of the original mixture?

A: If you use twice the volume of each of the solution to react, the amount of heat evolved would also be two times more. But this does not mean that the maximum temperature rise would be doubled as the amount of solution to "shoulder" the heat has also increased by two times. Thus, if you use the 'heat change = $V \cdot c \cdot \Delta T$' formula to calculate, ΔT is the same as the heat change doubles because V is doubled.

Q Is an indicator used in a normal acid–base titration better than in a thermometric titration?

A: Yes, it is more convenient and faster as you do not need to stop and wait to record the maximum rise in temperature after each addition of the titrant. Nevertheless, such planning allows us to learn the various possible ways of solving the same problem.

4.8 Determine the Purity of a Solid Through Thermometric Titration

The Task:

A student was given an impure sample of solid MgO. The percentage of MgO in the sample is approximately 98%. The reaction of MgO with HCl solution is an exothermic reaction:

$$\text{MgO(s)} + 2\text{HCl(aq)} \rightarrow \text{MgCl}_2\text{(aq)} + \text{H}_2\text{O(l)}, \quad \Delta H = \text{negative value.}$$

You are supposed to plan an experiment using thermometric titration, i.e., measuring the temperature of the mixture while adding different amounts of the solid sample, to determine the actual percentage purity of the solid MgO sample. You are provided with the following chemicals and apparatus:

- Some impure solid MgO;
- 1.00 mol dm^{-3} HCl solution;
- Styrofoam cup;
- Thermometer; and
- Standard glassware for titration.

> **Q** Where should we start thinking?

A: (i) What is the purpose of the plan?
— To determine the percentage purity of MgO in the impure sample.

(ii) What do you need to know in order to determine the percentage purity?
— We need to know the number of moles of MgO present in a fixed mass of the sample that we have measured.

(iii) How would you then determine the number of moles of MgO?
— We need to know the end-point of the titration between the HCl solution and solid MgO.

(iv) How do you get the end-point of the titration?
— Since the reaction between the acid and solid MgO is exothermic, as we add more solid MgO to a fixed volume of the acid, more heat energy is given off. Thus, the temperature of the solution is going to increase. At the end-point, a stoichiometric amount of MgO would have been added, meaning that the temperature rise is going to be at the maximum. After the end-point, if we continue to add solid MgO, the temperature of the solution is not going to change any further. Hence, from the intersection point, we can determine the number of moles of HCl needed and deduce the number of moles of MgO present in accordance to the following equation:

$$MgO(s) + 2HCl(aq) \rightarrow MgCl_2(aq) + H_2O(l),$$
$$\Delta H = \text{negative value}.$$

The graphical plot for the titration experiment is shown below:

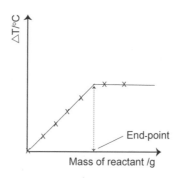

Pre-Experimental Calculations:

If we use 50.0 cm^3 of 1.00 mol dm^{-3} HCl solution:
$$MgO(s) + 2HCl(aq) \rightarrow MgCl_2(aq) + H_2O(l).$$
Amount of HCl present = $\frac{50}{1000} \times 1.00 = 5.0 \times 10^{-2}$ mol

Amount of MgO needed = $\frac{1}{2} \times$ Amount of HCl present = 2.5×10^{-2} mol

Molar mass of MgO = 24.3 + 16.0 = 40.3 g mol^{-1}

Mass of MgO needed = $40.3 \times 2.5 \times 10^{-2} = 1.00$ g

The range of the mass of MgO needed for the various experiments is as follows:

Vol. of HCl/cm^3	Intended mass of impure MgO/g	Actual mass of impure MgO/g	$\Delta T/°C$
50.0	0.20		
50.0	0.40		
50.0	0.60		
50.0	0.80		
50.0	1.00		
50.0	1.20		
50.0	1.40		
50.0	1.60		

The Procedure:

Procedure for measuring temperature rise:

(i) Set up the apparatus as shown in the diagram (refer to p. 104).
(ii) Use a 50 cm^3 *burette* to introduce *50.0 cm^3* of the acid solution into a *clean and dry styrofoam cup*.
(iii) Place the *thermometer* into the solution and take note of the *initial temperature* of the solution after some time.
(iv) *Weigh* accurately *0.20 g* of the impure sample and put it into the cup.

(v) *Stir gently* with the thermometer. Take note of the *highest rise* in temperature.
(vi) *Reweigh* the *emptied weighing bottle* to get the actual mass of the impure MgO that has been added.
(vii) *Repeat steps (ii) to (vi)* for different masses of the sample used.
(viii) *Plot a graph* of ΔT versus actual mass of impure sample added. The *intersection point* will give us the *mass of the impure sample* needed to completely react with the acid.
(ix) *Repeat the experiment* to get *reliable results*.

Q How do you calculate the percentage by mass of the MgO in the sample?

A: The mass of pure MgO needed to react with 50 cm^3 of 1.00 mol dm^{-3} HCl solution is 1 g. Hence,

percentage by mass of MgO

$$= \frac{1.00}{\text{Mass of impure sample obtained from the graph}} \times 100\%.$$

Note: In addition, such a planning process can also be used for the following similar cases:

(i) Determine the enthalpy change of reaction between zinc and copper(II) sulfate:

The Task:

Zinc reacts with copper(II) ion as follows:

$$Zn(s) + CuSO_4(aq) \rightarrow ZnSO_4(aq) + Cu(s).$$

You are supposed to plan an experiment using thermometric titration, i.e., measuring the temperature of the mixture while adding different amounts of the solid zinc, to determine the enthalpy change of reaction for the above equation. You are provided with the following chemicals and apparatus:

- Some solid zinc powder;
- 1.00 mol dm^{-3} CuSO$_4$ solution;
- Styrofoam cup;
- Thermometer; and
- Standard glassware for titration.

(ii) Determine the percentage by mass of KNO_3 in a sample containing KNO_3 and compound **X**:

> *The Task:*
>
> Given that the enthalpy change of solution of solid KNO_3 in water is endothermic in nature:
>
> $$KNO_3(s) + aq \rightarrow KNO_3(aq), \quad \Delta H = \text{positive value.}$$
>
> The dissolution of compound **X** in water does not result in any energy change. You are supposed to plan a thermometry experiment to determine the percentage by mass of KNO_3 in a sample containing KNO_3 and compound **X**.

The Approach:

(1) Measure the ΔT values for different masses of pure KNO_3 that has been dissolved in a fixed volume of pure water:

Vol. of water/cm^3	Mass of pure KNO_3/g	ΔT/°C
50.0	0.20	
50.0	0.40	
50.0	0.60	
50.0	0.80	
50.0	1.00	
50.0	1.20	
50.0	1.40	
50.0	1.60	

Plot a calibration graph of ΔT versus mass of KNO_3 used.

(2) Dissolve a known mass of the impure sample in the same volume of water. Measure the ΔT for the dissolving of this impure sample.
(3) From the calibration graph, find the actual mass of KNO_3 that gives rise to the ΔT value that has been measured in step (2).

4.9 Determine the Concentration of an Unknown Base Through Thermometry

The Task:

A student was given a monoacidic strong base solution (AOH) of concentration approximately 1 mol dm^{-3}. The reaction of the strong base with sulfuric acid solution is exothermic in nature. You are supposed to carry out a series of experiments, where different volumes of the AOH solution and H_2SO_4 solution are mixed and reacted. The temperature change, ΔT, is then plotted against the volume of H_2SO_4 solution used. Your plan should make use of the following chemicals and apparatus:

- AOH of unknown concentration;
- 0.500 mol dm^{-3} H_2SO_4 solution;
- Styrofoam cup;
- Thermometer; and
- Standard glassware in the lab.

Q Where should we start thinking?

A: (i) What is the purpose of the plan?
— To determine the concentration of the base, AOH.
(ii) What do you need to know in order to determine the concentration of the base, AOH?
— We need to know the number of moles of AOH present in a fixed volume of the sample solution that we have used.

(iii) How would you then determine the number of moles of AOH?

— Since the reaction between the base and H_2SO_4 is exothermic, as we add different volumes of the base and H_2SO_4 together, the reactions would release different amounts of heat energy. Now, if the total volumes of the mixtures are always kept at 100 cm^3, the temperature differences (ΔT) recorded are not going to be the same. If we plot a graph of ΔT versus the volume of H_2SO_4 solution added, we will get a "volcano" plot. The ΔT at the maximum point corresponds to the addition of a stoichiometric amount of the base and acid. Hence, from the intersection point, we can determine the number of moles of H_2SO_4 needed and deduce the number of moles of AOH present in accordance to the following equation:

$$H_2SO_4(aq) + 2AOH(aq) \rightarrow A_2SO_4(aq) + 2H_2O(l),$$
$$\Delta H = \text{negative value}.$$

The graphical plot for an exothermic and endothermic reaction are very similar to those in Section 4.7.

The Procedure:

Procedure for measuring ΔT:

(i) Set up the apparatus as shown in the diagram (refer to p. 104).
(ii) Use a *measuring cylinder* to introduce *10.0 cm^3* of the AOH solution into a *clean and dry styrofoam cup*.
(iii) Place the *thermometer* into the solution and take note of the *initial temperature* of the solution after some time.
(iv) Use another *measuring cylinder* to introduce *90.0 cm^3 of 0.500 mol dm^{-3} H_2SO_4* solution.
(v) *Stir gently* with the thermometer. Take note of the *highest rise* in temperature.
(vi) *Repeat steps (ii) to (v)* with *different volumes of AOH and H_2SO_4 solutions*, each time *maintaining a total volume of 100 cm^3*. Refer to the table below.
(vii) *Plot a graph* of temperature difference versus volume of H_2SO_4 solution added. The *intersection point* will give the *volume of H_2SO_4 needed to completely react with the base*.
(viii) *Repeat the experiment* to get *reliable results*.

Vol. of AOH/cm³	Vol. of H₂SO₄/cm³	Total Vol./cm³	$\Delta T/°C$
10.0	90.0	100.0	
20.0	80.0	100.0	
30.0	70.0	100.0	
40.0	60.0	100.0	
50.0	50.0	100.0	
60.0	40.0	100.0	
70.0	30.0	100.0	
80.0	20.0	100.0	
90.0	10.0	100.0	

Q Is the above experiment similar to the thermometric titration discussed in Section 4.7?

A: There is a major difference. The thermometric titration in Section 4.7 made use of a constant volume of analyte solution. As the titration progresses, the total volume of the mixture solution increases. But here in Section 4.9, the total volume of the mixture solution is always a constant. The key similarity between these two experiments is that the two graphical plots look similar. This is because we are finding an intersection point corresponding to the stoichiometric reaction between the two reactants that thus release the maximum amount of heat energy, for an exothermic reaction.

Q Why are measuring cylinders used to measure the volumes of the two solutions? Shouldn't a pipette or burette be used to measure the volume of the first solution?

A: As we are plotting a graph with different mixtures and we are only interested to find an intersection point on the graph, the usage of measuring cylinders is good enough. Even if there is a systematic error in the experiment, it would not affect the intersection point too badly. Of course, if you prefer to use a burette to measure the volume of the first solution to be introduced into the styrofoam cup, it is alright to do that.

4.10 Determine Whether the Acid Is a Strong or Weak Acid

The Task:

The enthalpy change of neutralization between a strong monobasic acid and strong monoacidic base to produce one mole of water is given as follows:

$$H^+(aq) + OH^-(aq) \rightarrow H_2O(l), \quad \Delta H = -57.0 \text{ kJ mol}^{-1}.$$

For weak acid–strong base, weak base–strong acid or weak base–weak acid neutralization, the enthalpy change is less exothermic than a strong acid–strong base reaction because the energy that is given off during neutralization has been partly diverted to help dissociate the undissociated weak acid or weak base.

A student was given a monobasic strong acid solution (HX) and a monobasic weak acid (HY), each of concentration 1.00 mol dm^{-3}. You are supposed to plan an experiment by determining which of the acid solutions is a strong acid, through thermochemical means. Your plan should make use of the following chemicals and apparatus:

- Solution A of concentration 1.00 mol dm^{-3};
- Solution B of concentration 1.00 mol dm^{-3};
- 1.00 mol dm^{-3} NaOH solution;
- Styrofoam cup;
- Thermometer; and
- Standard glassware in the lab.

Q Where should we start thinking?

A: (i) What is the purpose of the plan?

— To determine whether the acid solution is a strong or weak acid.

(ii) What do you need to know in order to determine whether the solution is a strong or weak acid?

— As each possible acid has its own strength, if we let a particular amount of NaOH to react with a particular volume of the acid, there would be a particular amount of heat energy evolved. This amount of heat energy evolved would give us a particular ΔT value. A stronger acid reacting with the strong base would give us more heat energy as compared to the reaction of the weak acid and the strong base. Thus, if we now compare this ΔT value with the expected ΔT value for each of the possible acids, we would be able to determine the strength of the acid. Or simply, the strong acid–strong base reaction would give us a greater ΔT value than a weak acid–strong base reaction.

Pre-Experimental Calculations:

Since the concentration of the acid is 1.00 mol dm^{-3}, if we each take 25.0 cm^3 of the acid:

Amount of HX/HY present = $\frac{25}{1000} \times 1.00 = 2.50 \times 10^{-2}$ mol

Thus, if we add 2.50×10^{-2} mol of 1.00 mol dm^{-3} NaOH into 25.0 cm^3 of the unknown acid solution:

$$H^+(aq) + OH^-(aq) \rightarrow H_2O(l).$$

Volume of NaOH solution = $\frac{2.50 \times 10^{-2}}{1.00}$ = 0.025 dm^3 = 25 cm^3

Total volume of the solution = 25.0 + 25.0 = 50.0 cm^3

Heat capacity of the solution is 4.18 J g^{-1} K^{-1} or 4.18 J cm^{-3} K^{-1}

For HX, the amount of H$_2$O formed = 2.50×10^{-2} mol

Assuming that HX is the strong acid while HY is the weak acid:

Enthalpy change, $\Delta H = \dfrac{\text{Heat evolved}}{\text{Amount of water formed in moles}} = -57.0$ kJ mol^{-1}

Heat evolved = $57.0 \times 2.50 \times 10^{-2}$ = 1.425 kJ = 1425 J

$$\Delta T = \frac{\text{Heat evolved}}{V \times c} = \frac{1425}{50 \times 4.18} = 6.8°C.$$

For HY, the ΔT is smaller than 6.8°C.

Hence, by measuring the ΔT for the reaction, we can determine the strength of the acid!

The Procedure:

Procedure for measuring ΔT:

(i) Set up the apparatus as shown in the diagram (refer to p. 104).
(ii) Use a *pipette* to introduce *25.0* cm^3 of the HX solution into a *clean and dry styrofoam cup*.
(iii) Place the *thermometer* into the solution and take note of the *initial temperature* of the solution after some time.
(iv) Use a *measuring cylinder* to introduce *25* cm^3 of 1.00 mol dm^{-3} of sodium hydroxide solution into the styrofoam cup.
(v) *Stir gently* with the thermometer.
(vi) Take note of the *highest rise* in temperature.
(vii) *Repeat steps (ii) to (vi)* for HY acid.
(viii) *Calculate the ΔT* and see whether the value is close to 6.8°C (strong acid) or smaller than 6.8°C (weak acid).
(ix) *Repeat the experiment* to get *reliable results*.

Note: In addition, such a planning process can also be used for the following similar cases:

(i) Determine the strength of a base using thermometry.

The Task:

The enthalpy change of neutralization between a strong monobasic acid and strong monoacidic base to produce one mole of water is given as follows:

$$H^+(aq) + OH^-(aq) \rightarrow H_2O(l), \quad \Delta H = -57.0 \text{ kJ mol}^{-1}.$$

For weak acid–strong base, weak base–strong acid or weak base–weak acid neutralization, the enthalpy change is less exothermic than a strong acid–strong base reaction because the energy given off during neutralization has been partly diverted to help dissociate the undissociated weak acid or weak base.

A student was given a monoacidic strong base solution (AOH) and a monoacidic weak base (BOH), each of concentration 1.00 mol dm^{-3}. You are supposed to plan an experiment by determining which of the monoacidic base solutions is a strong base, through thermochemical means. Your plan should make use of the following chemicals and apparatus:

- Solution X of concentration 1.00 mol dm^{-3};
- Solution Y of concentration 1.00 mol dm^{-3};
- 1.00 mol dm^{-3} HCl solution;
- Styrofoam cup;
- Thermometer; and
- Standard glassware in the lab.

(ii) Determine whether the acid is HCl or H_2SO_4.

The Task:

You are given two acids, HCl and H_2SO_4, each of the same concentration. Devise a thermometric plan to differentiate these two acids using a given standard sodium hydroxide solution.

(Note: The idea to solve this question is easy. The solution that contains H_2SO_4 would have twice the amount of H^+ ions than that for HCl. So, all you need to do is to add the same volume of NaOH solution to both the acid solutions. But the volume of the NaOH solution must be twice that of the acid solution. This is to ensure that all the NaOH have been reacted. The mixture that gives you a ΔT twice that of the other must be the H_2SO_4 solution.)

4.11 Determine the Enthalpy Change of Combustion of an Organic Compound

The Task:

Enthalpy change of combustion is the amount of energy that is evolved when one mole of substance is completely burned in excess oxygen. You are to devise a plan to determine the enthalpy change of combustion for liquid hexane. Your plan should make use of the following chemicals and apparatus:

- Hexane;
- Spirit lamp;

- Copper calorimeter;
- Thermometer; and
- Standard glassware in the lab.

Q Where should we start thinking?

A: (i) What is the purpose of the plan?
— To determine the enthalpy change of combustion of hexane.

(ii) What do you need to know in order to determine the enthalpy change of combustion of hexane?
— We need to know the amount of heat energy evolved and the amount of hexane being combusted, as the enthalpy change of combustion is calculated as follows:

Enthalpy change of combustion
$$= \frac{\text{Heat evolved}}{\text{Amount of compound combusted in moles}}.$$

(iii) How are you going to determine the amount of heat energy that has evolved?
— We can let the hexane burn and then "direct" the heat evolved to the copper calorimeter which contains about 100 cm^3 of water. By knowing the ΔT for the water and the heat capacity of the water, we can make use of $m \cdot c \cdot \Delta T$ to calculate the amount of heat energy that is "captured" by the water.

(iv) What would be a reasonable ΔT to measure?
— A ΔT of about 5°C is reasonable as too large a value would mean too much of the fuel needs to be burned. In addition, too long an experimental time may encourage more heat to be lost.

(v) How can you determine the amount of fuel that has burned?
— We can measure the weight of the spirit lamp before and after the burning; the difference would give us the mass of the fuel that has burned. Hence, if we know the molar mass of the fuel, we would be able to calculate the number of moles of fuel that has burned.

The Procedure:

Procedure for measuring ΔT:

(i) Set up the apparatus as shown in the diagram below.
(ii) Use a *measuring cylinder* to introduce *100* cm^3 of water into the *copper calorimeter*.
(iii) Place the *thermometer* into the water and take note of the *initial temperature* of the water after some time.
(iv) Place some *hexane* into the *spirit lamp*. Put a *wick* onto it. Take the *mass of the spirit lamp*.
(v) Place the spirit lamp *under the calorimeter* and *light up* the spirit lamp.
(vi) *Stir the water gently* with the thermometer *continuously*. Do not let the thermometer touch the base of the copper calorimeter.
(vii) When the temperature of the water has *increased by about 5°C*, *extinguish* the spirit lamp by blowing the flame out. Take the *mass of the spirit lamp* immediately.
(viii) Read the *highest rise* in temperature and *calculate the* ΔT.
(ix) Calculate the *mass of the hexane used*.
(x) *Repeat the experiment* to get *reliable results*.

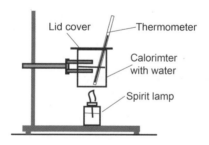

Q Why is a copper calorimeter being used? Can we use a normal beaker?

A: A copper calorimeter is preferred to a normal glass beaker because copper is a very good thermal conductor. Remember, we need to "capture" as much heat energy into the water as possible.

Planning for Energetics Experiments 143

 So, does that mean that all the heat energy that is evolved from the combustion goes into the water?

A: Of course not! Some of the heat energy has been absorbed by the copper calorimeter itself while some of heat energy is lost due to convection.

 What happen if the flame of the spirit lamp is too far away from the calorimeter?

A: If the flame tip is too far away from the calorimeter, less heat energy would reach the calorimeter. This would mean more heat energy would be lost. Thus, the distance of the flame to the calorimeter is very important.

 Why can't we let the thermometer touch the base of the copper calorimeter?

A: As the base of the calorimeter comes in contact with the heat source directly, its temperature is going to be higher than that of the water.

 Why is the mass of the spirit lamp taken immediately after the flame has been extinguished? Shouldn't we wait for the lamp to cool down first?

A: If you wait for the lamp to cool down, the fuel, which is volatile, will be lost due to evaporation. Hence, the mass reading would be inaccurate. So, it is better to take the mass of the lamp immediately after the flame has been extinguished. You would then need to know when and why to do the right thing. Sometimes there is trade-off for doing the right thing.

Result and Calculations:

Mass of spirit lamp and hexane before combustion/g	X
Mass of spirit lamp and hexane after combustion/g	Y

Temperature of water in copper calorimeter before combustion/°C	A
Temperature of water in copper calorimeter after combustion/°C	B

Mass of hexane combusted = (X − Y) = W g

Amount of hexane combusted = $\dfrac{W}{\text{Molar mass of hexane}}$ = Z mol

ΔT = (B − A) = D °C

Volume of water used = 100 cm^3

Assume density of water = 1.0 g cm^{-3}

Heat capacity of water = 4.18 J g^{-1} K^{-1}

Assume no heat loss to the surroundings:

Heat evolved from combustion = Heat absorbed by water
$\quad\quad\quad\quad\quad\quad\quad\quad\quad\quad\quad\quad$ = $\rho \cdot V \cdot c \cdot \Delta T$
$\quad\quad\quad\quad\quad\quad\quad\quad\quad\quad\quad\quad$ = 100 × 4.18 × D = 418D J

Hence, enthalpy change of combustion of hexane = $-\dfrac{418D}{Z}$ J mol^{-1}.

Q: Since the copper calorimeter actually did absorb some heat, is there a way to calculate the amount of heat that is taken in by the copper calorimeter?

A: Oh yes. You can burn some fuel with a known enthalpy change of combustion and use this to calculate the heat capacity of the calorimeter. So, once you know the heat capacity of the copper calorimeter, you can use it to calculate the amount of heat energy it will absorb in the future when you use it. Refer to the next section to see how we can do it.

Note: In addition, such a planning process can also be used for the following similar cases:

(i) Determine the enthalpy change of combustion for methanol, ethanol, propan-1-ol, butan-1-ol, pentan-1-ol, and hexan-1-ol. Plot a graph of the enthalpy change of combustion for these alcohols using these alcohols as the *x*-axis.

— For such a planning exercise, it is important to mention that the wick needs to be changed before starting the combustion experiment for the next alcohol. This is because if the wick is not changed, it would be soaked with the previous alcohol. Then the experiment would be inaccurate! In addition, the tip of the flame must be at the same distance from the bottom of the calorimeter. This is to ensure that relatively similar amounts of heat energy are being lost because of this factor.

(ii) Determine the enthalpy change of combustion of an ethanol–propanol mixture.
— For such a planning exercise, just treat the fuel mixture as a single component. Thus, there is nothing to worry about as the big concepts are all similar.

4.12 Determine the Heat Capacity of a Calorimeter

The Task:

The enthalpy change of combustion of hexane is -4160 kJ mol^{-1}. Use this value to plan an experiment to determine the heat capacity of a copper calorimeter. Your plan should make use of the following chemicals and apparatus:

- Hexane;
- Spirit lamp;
- Copper calorimeter of mass, m g;
- Thermometer; and
- Standard glassware in the lab.

Q Where should we start thinking?

A: (i) What is the purpose of the plan?
— To determine the heat capacity of a copper calorimeter.
(ii) What do you need to know in order to determine the heat capacity of a copper calorimeter?
— We need to know the amount of heat energy absorbed by the copper calorimeter as the heat energy absorbed is calculated as follows:
Heat absorbed by calorimeter = $m \cdot C \cdot \Delta T$, where C is the heat capacity.

(iii) How can you determine the amount of heat energy that has been absorbed?
— We can let the hexane burn and "direct" the heat energy to the copper calorimeter which contains about 100 cm^3 of water. By knowing the ΔT for the water and the heat capacity, we can make use of $V \cdot c \cdot \Delta T$ and $m \cdot C \cdot \Delta T$ to calculate the amount of heat energy that is "captured" by the water and the calorimeter, respectively.

(iv) What would be a reasonable ΔT to measure?
— A ΔT of about 5°C is reasonable as too large a value would mean that too much of the fuel needs to be burned. In addition, too long an experimental time may encourage more heat to be lost.

(v) How can you determine the amount of fuel that has burned?
— We can measure the weight of the spirit lamp before and after burning; the difference would give us the mass of the fuel that has burned. Hence, if we know the molar mass of the fuel, we would be able to calculate the number of moles of fuel that has burned.

(vi) How can you determine the amount of heat energy that has evolved?
— Well, if we know both the amount of hexane that has burned and its enthalpy change of combustion, we can obtain the amount of heat energy that has evolved.

The Procedure:

Procedure for measuring ΔT:

(i) Set up the apparatus as shown in the diagram in Section 4.11.
(ii) Use a *measuring cylinder* to introduce *100* cm^3 of water into the *copper calorimeter*.
(iii) Place the *thermometer* into the water and take note of the *initial temperature* of the water after some time.
(iv) Place some *hexane* into the *spirit lamp*. Put a *wick* onto it. Take the *mass of the spirit lamp*.
(v) Place the spirit lamp *under the calorimeter* and *light up* the spirit lamp.
(vi) *Stir the water gently* with the thermometer. Do not let the thermometer touch the base of the copper calorimeter.

(vii) When the temperature of the water has *increased by about 5°C*, *extinguish* the spirit lamp. Take the *mass of the spirit lamp* immediately.
(viii) Read the *highest rise* in temperature and *calculate the* ΔT.
(ix) Calculate the *mass of hexane used*.
(x) *Repeat the experiment* to get *reliable results*.

Result and Calculations:

Mass of spirit lamp and hexane before combustion/g	X
Mass of spirit lamp and hexane after combustion/g	Y

Temperature of water in copper calorimeter before combustion/°C	A
Temperature of water in copper calorimeter after combustion/°C	B

Mass of hexane combusted = (X − Y) = W g

Amount of hexane combusted = $\dfrac{W}{\text{Molar mass of hexane}}$ = Z mol

Given that the enthalpy change of combustion of hexane is −4160 kJ mol^{-1}, therefore the amount of heat evolved = Z × 4160 = 4160Z kJ

$\Delta T = (B - A) = D°C$

Volume of water used = 100 cm^3

Assume density of water = 1.0 g cm^{-3}

Heat capacity of water = 4.18 J g^{-1} K^{-1} = 0.00418 kJ g^{-1} K^{-1}

Assume that there is no heat loss to the surroundings:

Heat evolved = Heat absorbed by water + Heat absorbed by copper calorimeter

$\Rightarrow$ 4160Z = $\rho \cdot V \cdot c \cdot \Delta T + m \cdot C \cdot \Delta T$
$\Rightarrow$ 4160Z = 100 × 0.00418 × D + m × C × D
$\Rightarrow$ 4160Z = 0.418D + $m \cdot C \cdot D$

Therefore, the heat capacity of the copper calorimeter, $C = \dfrac{(4160Z - 0.418D)}{m.D}$ kJ g^{-1} K^{-1}.

4.13 Determine the Enthalpy Change of Combustion of Hexane Taking into Consideration the Heat Absorbed by the Calorimeter

The Task:

Enthalpy change of combustion is the amount of heat energy evolved when one mole of substance is completely burned in excess oxygen. You are to devise a plan to determine the enthalpy change of combustion for liquid hexane. Your plan should make use of the following chemicals and apparatus:

- Hexane;
- Spirit lamp;
- Copper calorimeter of heat capacity, C J K^{-1};
- Thermometer; and
- Standard glassware in the lab.

Q Where should we start thinking?

A: (i) What is the purpose of the plan?

— To determine the enthalpy change of combustion of hexane.

(ii) What do you need to know in order to determine the enthalpy change of combustion of hexane?

— We need to know the amount of heat energy evolved and the amount of hexane being combusted as the enthalpy change of combustion is calculated as follows:

Enthalpy change of combustion

$$= \frac{\text{Heat evolved}}{\text{Amount of compound combusted in moles}}.$$

(iii) How can we determine the amount of heat energy evolved?

— We can let the hexane burn and "direct" the heat energy to the copper calorimeter which contains about 100 cm^3 of water. By knowing the ΔT for the water and the heat capacity, we can make use of $V \cdot c \cdot \Delta T$ to calculate the heat energy that is "captured" by the water.

And by knowing the heat capacity of the calorimeter, C J K^{-1}, we can also calculate the amount of heat energy that has been absorbed by the calorimeter.

(iv) What would be a reasonable ΔT to measure?

— A ΔT of about 5°C is reasonable as too large a value would mean too much of the fuel needs to be burned. In addition, too long an experimental time may encourage more heat to be lost.

(v) How can you determine the amount of fuel that has burned?

— We can measure the weight of the spirit lamp before and after the burning; the difference would give us the mass of the fuel that has burnt. Hence, if we know the molar mass of the fuel, we would be able to calculate the number of moles of fuel that has burnt.

The Procedure:

Procedure for measuring ΔT:

(i) Set up the apparatus as shown in the diagram in Section 4.11.
(ii) Use a *measuring cylinder* to introduce *100*cm^3 into the *copper calorimeter*.
(iii) Place the *thermometer* into the water and take note of the *initial temperature* of the water after some time.
(iv) Place some *hexane* into the *spirit lamp*. Put a *wick* onto it. Take the *mass of the spirit lamp*.
(v) Place the spirit lamp *under the calorimeter* and *light up* the spirit lamp.
(vi) *Stir the water gently* with the thermometer. Do not let the thermometer touch the base of the copper calorimeter.
(vii) When the temperature of the water has *increased by about 5°C*, *extinguish* the spirit lamp. Take the *mass of the spirit lamp* immediately.
(viii) Read the *highest rise* in temperature and *calculate the ΔT*.
(ix) Calculate the *mass of hexane used*.
(x) *Repeat the experiment* to get *reliable results*.

Result and Calculations:

Mass of spirit lamp and hexane before combustion/g	X
Mass of spirit lamp and hexane after combustion/g	Y

Temperature of water in copper calorimeter before combustion/°C	A
Temperature of water in copper calorimeter after combustion/°C	B

Mass of hexane combusted = (X − Y) = W g

Amount of hexane combusted = $\dfrac{W}{\text{Molar mass of hexane}}$ = Z mol

ΔT = (B − A) = D °C

Volume of water used = 100 cm^3

Assume density of water = 1.0 g cm^{-3}

Heat capacity of water = 4.18 J g^{-1} K^{-1}

Heat capacity of calorimeter = C J K^{-1}

Assume that no heat loss to the surroundings:

Heat evolved = Heat absorbed by water + Heat absorbed by copper calorimeter

$\quad\quad\quad\quad\;\; = \rho \cdot V \cdot c \cdot \Delta T + C \cdot \Delta T$

$\quad\quad\quad\quad\;\; = 100 \times 4.18 \times D + C \times D$

$\quad\quad\quad\quad\;\; = (418D + C \cdot D)$ J

Hence, enthalpy change of combustion of hexane = $-\dfrac{(418D + C \cdot D)}{Z}$ J mol^{-1}.
(Take note of the differences between Sections 4.11, 4.12, and 4.13).

4.14 Determine the Enthalpy Change of Hydrogenation of Pent-1-yne

The Task:

Pent-1-yne, $CH_3CH_2CH_2C\equiv CH$, undergoes the following hydrogenation to form pent-1-ene:

$\quad CH_3CH_2CH_2C\equiv CH(l) + H_2(g) \rightarrow CH_3CH_2CH_2CH=CH_2(l).$

Given the enthalpy change of combustion of hydrogen gas is -286 kJ mol^{-1}, devise a plan to determine the enthalpy change of hydrogenation of pent-1-yne. Your plan should make use of the following chemicals and apparatus:

- Pent-1-yne;
- Pen-1-ene;
- Spirit lamp;
- Copper calorimeter;
- Thermometer; and
- Standard glassware in the lab.

Q Where should we start thinking?

A: (i) What is the purpose of the plan?
— To determine the enthalpy change of hydrogenation of pent-1-yne.

(ii) What do we need to know in order to determine the enthalpy change of hydrogenation of pent-1-yne?
— First, we need to determine the enthalpy change of combustion of pent-1-yne and pent-1-ene, then construct an energy cycle to calculate the enthalpy change of hydrogenation of pent-1-yne:

$$CH_3CH_2CH_2C\equiv CH(l) + H_2(g) \xrightarrow{\Delta H_{hydrogenation}} CH_3CH_2CH_2CH=CH_2(l)$$

with $+0.5\ O_2(g)$, $\Delta H_c[H_2(g)]$ going down to $CH_3CH_2CH_2C\equiv CH(l) + H_2O(l)$, then $\Delta H_c[\text{pent-1-yne}]$ with $+7\ O_2(g)$ to $5CO_2(g) + 5H_2O(l)$; and $+7.5\ O_2(g)$, $\Delta H_c[\text{pent-1-ene}]$ from the top right down to $5CO_2(g) + 5H_2O(l)$.

By Hess's Law, $\Delta H_{\text{hydrogenation}} = \Delta H_C[H_2(g)] + \Delta H_C[\text{pent-1-yne}] - \Delta H_C[\text{pent-1-ene}]$.

152 *Understanding Experimental Planning for Advanced Level Chemistry*

(iii) How can we determine the enthalpy change of combustion?

— Well, we can refer to Section 4.11.

Note: In addition, such a planning process can also be used for the following similar cases:

The Task:

Determine the enthalpy change of isomerization of *cis*-pent-2-ene to *trans*-pent-2-ene by determining the enthalpy changes of combustion of the two alkenes:

$$cis\text{-}CH_3CH_2CH=CHCH_3(l) \xrightarrow{\Delta H_{isomerization}} trans\text{-}CH_3CH_2CH=CHCH_3(l)$$

$\Delta H_c[cis\text{-pent-1-ene})]$ $\Delta H_c[trans\text{-pent-1-ene})]$

$+7.5\ O_2(g)$ $+7.5\ O_2(g)$

$$5CO_2(g) + 5H_2O(l)$$

By Hess's Law, $\Delta H_{isomerization} = \Delta H_C[cis\text{-pent-1-ene}] - \Delta H_C[trans\text{-pent-1-ene}]$.

4.15 Determine How Boiling Point Is Affected by the Presence of Impurities

The Task:

Boiling happens when the vapor pressure of the liquid is equal to that of the atmospheric pressure. The presence of impurities, such as NaCl, will decrease the vapor pressure of the liquid as the water molecules are strongly attracted to the ions via ion–dipole interactions. As such, a higher temperature is needed to create a vapor pressure that is high enough to overcome the atmospheric pressure. You are to plan an experiment to determine how the boiling point of water changes with an increase in the level of impurities. Your plan should make use of the following chemicals and apparatus:

- Solid sodium chloride;
- Deionized water;
- Thermometer; and
- Standard glassware in the lab.

Q Where should we start thinking?

A: (i) What is the purpose of the plan?

— To determine how the boiling point of water changes with an increase in the amount of NaCl.

(ii) How can we determine how the boiling point of water changes with an increase in the amount of NaCl?

— We can measure the boiling point of pure water and those in which some solid NaCl has been dissolved.

(iii) So, how would you know how much of the water to use and how much of the solid NaCl to add?

— Well, we can assume that we are going to use 50 cm³ of water and adding different amounts of NaCl of masses 1 g, 2 g, 3 g, 4 g, and 5 g. We can then measure the boiling point of each of these mixtures.

The Procedure:

(i) Use a 50 cm³ *burette* to introduce *50 cm³ of water* into a *clean and dry round-bottomed flask* as shown in the set-up below:

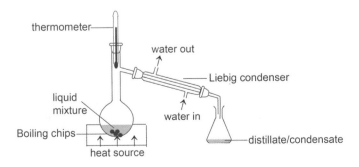

(ii) Introduce some *boiling chips* and start the heating process. Take note of the *temperature on the thermometer when the liquid boils*.
(iii) Discard the contents. *Repeat steps* (i) to (iii) by adding 1 g, 2 g, 3 g, 4 g, and 5 g of solid NaCl into 50 cm^3 of water.
(iv) *Repeat the experiment* to get *reliable results*.
(v) Plot a graph of boiling point versus mass of sodium chloride added.

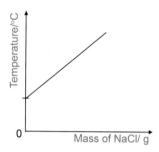

4.16 Safety Precautions for Thermometry Experiments

You may be asked to quote some safety precautions while performing a thermometry experiment. Depending on the type of thermometric experiments that you are performing, the following examples may be useful for you to take note:

— Always wear gloves, a lab coat, and safety goggles while performing experiments. For example, solid sodium hydroxide, acids (H$_2$SO$_4$) or bases (NaOH), methanol, etc. that you will be using may be corrosive in nature. Thus, there should be minimal direct contact of the skin with these chemicals.
— The styrofoam cup should be placed inside a beaker so that it does not topple over easily, causing the thermometer to break. The broken glasses can cut and the mercury is poisonous.
— If there are any toxic gases or fumes evolved, conduct the experiment in a fumehood as such fumes may cause respiratory problems.

- If there are flammable liquids used, such as alcohol, ensure there is no naked flame around.
- Any organic chemicals should be properly covered or capped when not in use and used organic waste must be properly disposed.

4.17 Minimizing Experimental Errors or Increasing Reliability

You may be asked to quote some ways to improve the experimental reliability while performing a thermometric experiment. Depending on the type of thermometric experiment that you are performing, the following examples may be useful for you to take note:

- Some of the compounds, such as the Group 1 hydroxides, are hydroscopic in nature. Weigh the compound quickly or cover it after weighing to avoid the absorption of moisture from the air.
- In the making of solution, if the compounds that dissolve give off or absorb heat energy, wait for the solution to reach thermal equilibrium with the surroundings before using them.
- As heat is easily lost to the surroundings through convection, perform a thermometric experiment in a draught-free environment.
- Cover the styrofoam cup with a lid or provide lagging to minimize heat loss.
- When conducting a combustion experiment, the metal calorimeter should be lagged on the sides, not at the bottom, which needs to be in contact with the flame.
- The flame cannot be too far away from the metal calorimeter as this would lead to more heat lost by convection.
- The accuracy of the mass of spirit lamp measured is compromised by the burning of the wick. This is unavoidable.
- If the volume of the solution needs to be more precise, use a pipette or burette instead of a measuring cylinder as the latter is a less-precise apparatus.
- Measure the initial temperatures of two solutions first before mixing. Calculate the weighted average temperature if necessary.

- Use a thermometer with a smaller division such as 0.5°C for measuring small temperature changes. Usage of larger-division thermometers for small temperature changes, such as 1°C, leads to a greater reading error.
- If a more accurate measurement of the temperature is required, use a data logger with a temperature sensor as the response of such instrument is better than that of a mercury thermometer.
- The experiment must always be repeated to check for reliability of results. The average value should be calculated if necessary.

CHAPTER 5

PLANNING FOR KINETICS EXPERIMENTS

Chemical kinetics is the study of the rate of a chemical reaction and the factors affecting the rate of the reaction. The rate of the reaction is defined as follows:

Rate of reaction = $\frac{\Delta c}{\Delta t}$ where Δc is the change of concentration of a reactant or product.

The unit or dimension for kinetics is mol dm^{-3} time^{-1}. Thus, from the unit, it is obvious that the rate can be determined from one of the following ways:

- The initial rate method — Measure the time taken (Δt) for a particular concentration change (Δc).
- The continuous method — Monitor the concentration of a species continuously with respect to time. Then, plot a graph of concentration versus time. The gradient $\left(\frac{dc}{dt}\right)$ of the plot at specific points of time gives us the rate.

Q How do you monitor the concentration of a species in a reaction?

A: There are numerous ways. Basically, it can be broadly classified into (1) chemical methods and (2) physical methods. Chemical methods refer to methods that make use of a chemical reaction, e.g. titration, to probe the concentration of the species at specific points of time during the reaction. So, such methods are "intrusive" in nature. Physical methods on the other hand, make use of monitoring a physical property of the reaction system, which can be pressure, mass, volume, temperature, color intensity, pH, electrical conductivity, etc. The concept behind physical methods is that the rate of change of such a physical property is proportional to the rate of the chemical reaction. This is preferred to the titrimetric method as it does not "disturb" the reaction. For more details, you can refer to *Understanding Advanced Physical Inorganic Chemistry* by J. Tan and K. S. Chan.

To study the kinetics of the reaction: aA + bB → cC + dD, i.e., to determine the rate equation, rate = $k[A]^n[B]^m$, where k is the rate constant. The dependent variable is the rate while the independent variable is either [A] or [B]. For example:

(1) Using the initial rate method
   — To determine the value of *n*: Perform two sets of experiments in which [B] are the same; only [A] changes. Measure how the rate changes with [A] being changed.
   — To determine the value of *m*: Perform two sets of experiments in which [A] are the same; only [B] changes. Measure how the rate changes with [B] being changed.

(2) Using the continuous method
   — To determine the value of *n*: Perform one set of experiment in which [B] is much higher than [A]. Measure [A] with respect to time.
   — To determine the value of *m*: Perform one set of experiment in which [A] is much higher than [B]. Measure [B] with respect to time.

Q What are the numbers, *n* and *m*, known as?

A: They are known as the order of the reaction. Basically, these numbers tell us the extent the rate of the reaction depends on the concentrations of each of the reactants.

Q What is the significance of keeping the concentration of some reactants higher than that of the independent variable?

A: When you perform an experiment, you can only change one variable while keeping the rest of the variables constant. In the continuous method, how do we ensure that the rest of the variables do not affect our independent variable? We make sure that the concentrations of the non-independent variables are higher than that of the independent variable. Why? As the reaction proceeds, the concentrations of both the independent and non-independent variables are going to decrease. But because the concentrations of the non-independent variables are higher, their decrease in concentrations would not affect the rate of the reaction significantly. That is, the concentrations of these non-independent variables are essentially constant relative to the decreasing concentration of the independent variable. Hence, we can ascribe the rate of the reaction as ONLY due to the decrease in the concentration of the independent variable.

> **Q** For the initial rate method, is it necessary to keep the concentration of some reactants higher than that of the independent variable?

A: Not really. This is because the time interval used to measure the rate of change using the initial rate method is usually much shorter than the continuous method. For such a small time interval, the change of concentration of the non-independent variables, do not affect the measurement of the rate significantly.

In addition, only the concentration of the independent variable is changed while the concentrations of the other reactants are kept constant between two different sets of experiment. Hence, the rate of the reaction can only be affected by the change in concentration of the independent variable.

> **Q** Using the continuous method, how can we determine the order of the reaction with respect to the reactant?

A: For a zero-order reaction, the rate equation is rate = $k[A]^0$ = k. Mathematically, the graphs would look as follows:

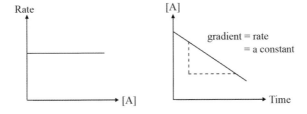

For a first-order reaction, the rate equation is rate = k[A]. Mathematically, the graphs would look as follows:

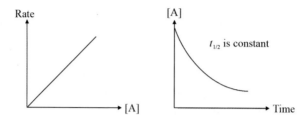

For a second-order reaction, the rate equation is rate = k[A]². Mathematically, the graphs would look as follows:

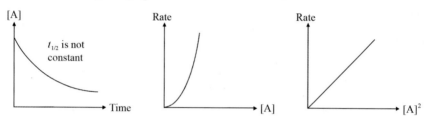

From the above, the shape of the concentration versus time plot would tell us if it is a zero-order reaction. Determining whether the curve has a constant half-life would inform us if it is a first-order reaction.

 By using the continuous method, how do we know the order of reaction with respect to [B] if only the concentration of A is "monitorable?"

A: Simple! Perform two experiments keeping [A] constant while changing [B]. These are the possible scenarios from the plot of [A] versus time graph:

(1) Reaction is zero-order for both [A] and [B]:

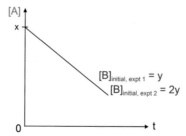

(2) Reaction is zero-order for [A] but not for [B]:

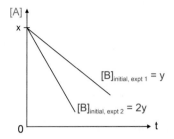

(3) Reaction is zero-order for [B] but not for [A]:

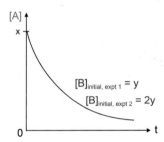

(4) Reaction is not zero-order for both [A] and [B]:

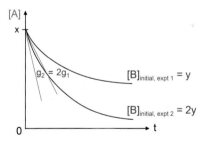

5.1 Determine the Rate of Reaction Between HCl and CaCO$_3$ by Gravimetry

The Task:

Calcium carbonate reacts with hydrochloric acid in accordance to the following equation:

$$CaCO_3(s) + 2HCl(aq) \rightarrow CaCl_2(aq) + CO_2(g) + H_2O(l).$$

You are supposed to plan an experiment to determine the rate of reaction for the above reaction with the following chemicals and apparatus:

- 5 g of solid CaCO$_3$;
- 1.50 mol dm^{-3} HCl solution;
- Weighing balance;
- Stopwatch; and
- Standard glassware in the lab.

Q Where should we start thinking?

A: (i) What is the purpose of the plan?
— To determine the rate of reaction between HCl and CaCO$_3$.
(ii) What do you need to know in order to determine the rate of reaction between HCl and CaCO$_3$?
— We need to know how fast the CaCO$_3$ has reacted with the HCl.
(iii) How do you then measure the rate?
— As the following reaction proceeds, CO$_2$ gas is produced. Thus, when the CO$_2$ gas is escaping, the mass of the system would decrease. Hence, we can use the weighing balance to monitor the mass of the system with respect to time:

$$CaCO_3(s) + 2HCl(aq) \rightarrow CaCl_2(aq) + CO_2(g) + H_2O(l).$$

The rate of change of the mass is proportional to the rate of the reaction.

(iv) How can you determine how much of the acid to use?

— Well, we can assume that a certain mass of $CaCO_3$ is used and then calculate the theoretical amount of HCl that is needed.

Pre-Experimental Calculations:

$$CaCO_3(s) + 2HCl(aq) \rightarrow CaCl_2(aq) + CO_2(g) + H_2O(l)$$

Assuming that about 2.5 g of $CaCO_3$ is used:

Molar mass of $CaCO_3 = 40.1 + 12.0 + 3(16.0) = 100.1 \text{ g mol}^{-1}$

Amount of $CaCO_3$ in 2.5 g $= \dfrac{2.5}{100.1} = 2.50 \times 10^{-2}$ mol

Amount of HCl needed $= 2 \times$ Amount of $CaCO_3$ in 2.5 g

$= 2 \times 2.50 \times 10^{-2} = 5.00 \times 10^{-2}$ mol

Volume of 1.50 mol dm^{-3} HCl solution $= \dfrac{5.00 \times 10^{-2}}{1.5} = 0.0333$ dm^3

$= 33.3$ cm^3.

The Procedure:

The general procedure of a kinetics experiment involves the measurement of concentration or some other variables that are related to the concentration values with respect to time. When a student follows the experimental procedure diligently, the student should be able to obtain repeatable results. Similarly, another student doing the same experiment should obtain data that are close to what the rest have obtained. Thus, the procedure of a typical kinetics experiment would consist of a series of steps which inform the student: (1) which step should come first; (2) what apparatus should he/she use; (3) what is the quantity of substance that he/she should measure; and (4) if needed, the reaction conditions such as temperature and pressure.

(i) Set up the apparatus as shown in the diagram below.

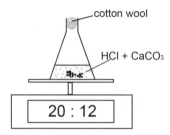

(ii) *Weigh* accurately *2.5 g of $CaCO_3$* into the *conical flask*.
(iii) Use a measuring cylinder to introduce *35 cm^3* of *1.50 mol dm^{-3} HCl solution* into the conical flask.
(iv) Stopper the mouth of the flask with some *cotton wool*. Quickly *start the stopwatch* and take the *mass reading at t = 0* min.
(v) Record the *mass reading of the set-up* at 30-s intervals until a *relatively constant mass is obtained*.
(vi) *Plot a graph* of mass versus time.
(vii) *Repeat the experiment* to get *reliable results*.

> **Q** What is the purpose of putting some cotton wool at the mouth of the conical flask?

A: As the CO_2 gas produced, it might create some acid spray which would result in the loss of material.

> **Q** Can we determine the rate of reaction by gas collection instead?

A: Yes, certainly. But do not collect CO_2 through the downward displacement of water as CO_2 gas is partially soluble in water.

Note: You can also use the same planning outline for:

Determine the rate of reaction of Group 2 carbonates ($MgCO_3$, $CaCO_3$, $SrCO_3$, $BaCO_3$) with acid down the group using the gas collection method.

(Note: For this planning exercise, make sure that the numbers of moles of the Group 2 carbonates that you use are the same. This would thus ensure a similar volume of CO_2 gas collected.)

5.2 Determine the Rate of Decomposition of H_2O_2 Using the Gas Collection Method

The Task:

Hydrogen peroxide decomposes slowly in accordance to the following equation:

$$2H_2O_2(aq) \rightarrow 2H_2O(l) + O_2(g).$$

It is found that the reaction is first-order with respect to H_2O_2. The reaction can be accelerated by using manganese(IV) oxide as the catalyst.

You are supposed to plan an experiment to determine the rate of reaction for the above reaction with the following chemicals and apparatus:

- 50 cm^3 of 0.150 mol dm^{-3} H_2O_2 solution;
- Solid manganese(IV) oxide;
- Gas syringe;
- Stopwatch; and
- Standard glassware in the lab.

Q Where should we start thinking?

A: (i) What is the purpose of the plan?
— To determine the rate of decomposition of H_2O_2.
(ii) What do you need to know in order to determine the rate of decomposition of H_2O_2?
— We need to know how fast the H_2O_2 has reacted with respect to time.

(iii) How do you then measure the rate?

— As the following reaction proceeds, O_2 gas is produced. Thus, we can measure the volume of the O_2 gas collected with respect to time:

$$2H_2O_2(aq) \rightarrow 2H_2O(l) + O_2(g).$$

The rate of the volume of the O_2 gas collected is proportional to the rate of the reaction.

(iv) But how can you determine how much of the H_2O_2 solution to use?

— We need to assume that the volume of gas collected is 50 cm^3. By making this assumption, we would know the maximum number of moles of gas particles to be collected and then deduce the amount of H_2O_2 needed to be decomposed.

(v) The decomposition of H_2O_2 is rather slow; how can you speed it up?

— Well, we can add some manganese(IV) oxide as a catalyst.

Pre-Experimental Calculations:

$$2H_2O_2(aq) \rightarrow 2H_2O(l) + O_2(g)$$

Assuming that 50 cm^3 of gaseous molecules is collected at r.t.p. conditions:

At r.t.p. (1 atm, 20°C), molar volume = 24.0 dm^3 mol^{-1}.

Amount of O_2 gas = $\dfrac{50}{24{,}000}$ = 2.08×10^{-3} mol

Amount of H_2O_2 = 2 × Amount of O_2 gas = 4.16×10^{-3} mol

Since the concentration of the H_2O_2 solution is 0.150 mol dm^{-3},

Volume of H_2O_2 solution needed = $\dfrac{4.16 \times 10^{-3}}{0.150}$ = 0.0277 dm^3 = 27.7 cm^3.

The Procedure:

(i) Set up the apparatus as shown below.

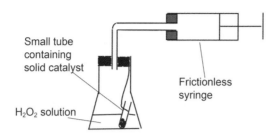

(ii) Use a 50 cm^3 *burette* to introduce *27.00* cm^3 of H$_2$O$_2$ solution into the conical flask.
(iii) Attached a *small tube containing some solid manganese(IV) oxide*.
(iv) Record the *initial volume reading* of the syringe.
(v) Let the *manganese(IV) oxide comes in contact with the solution* by loosening the stopper. *Start the stopwatch* immediately. Swirl to mix *gently*.
(vi) Record the *volume of the gas collected* at 30-s intervals until a *relatively constant volume* is obtained.
(vii) *Plot a graph* of volume of gas versus time.
(viii) *Repeat the experiment* to get *reliable results*.

Q How would the graphical plots for the results in Sections 5.1 and 5.2 look like?

A: When the experimental values of the amount and time are plotted graphically, we would obtain the following graphs:

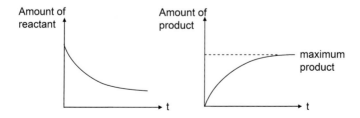

The shapes of the graphs are meant for a non-zero order reaction. If the reaction is zero-order with respect to the reactant that we are monitoring, the graph would be a straight-line plot with a constant gradient. The constant gradient means that the rate of the reaction is the same, irrespective of the concentration of the reactant in the system at any time t, i.e., rate = k, where k is the rate constant.

Note: You can also use the same planning outline for:

(i) Determine the rate of reaction of benzenediazonium ions with water at 25°C.

The Task:

At a temperature of about 10°C, benzenediazonium ions react with water slowly in accordance to the following equation to give phenol (C_6H_5OH) and nitrogen gas:

$$C_6H_5N_2^+(aq) + H_2O\ (l) \rightarrow C_6H_5OH(aq) + N_2(g) + H^+(aq).$$

But at room temperature of about 25°C, the reaction becomes significant. Design an experiment that will enable you to determine the order of reaction with respect to benzenediazonium ion at 25°C. You are provided with the following chemicals and apparatus:

- 0.150 mol dm^{-3} of aqueous benzenediazonium chloride solution at 10°C;
- Deionized water;
- Gas syringe;
- Stopwatch; and
- Standard glassware in the lab.

(Note: Here, the volume of the N_2 gas is collected and measured with respect to time.)

(ii) Determine the rate of reaction between sodium and ethanol.

> *The Task:*
>
> Ethanol is acidic in nature. Sodium being a reactive metal reacts with ethanol as follows:
> $$2CH_3CH_2OH + 2Na \rightarrow 2CH_3CH_2ONa + H_2.$$
> Design an experiment that will enable you to determine the order of reaction with respect to ethanol, using the gas collection method.

(Note: Here, the volume of the H_2 gas is collected and measured with respect to time.)

5.3 Determine the Rate of Decomposition of H_2O_2 Using Titration

> *The Task:*
>
> Hydrogen peroxide undergoes a redox reaction with acidified $KMnO_4$ as follows:
> $$2MnO_4^-(aq) + 5H_2O_2(aq) + 6H^+(aq) \rightarrow 2Mn^{2+}(aq) + 5O_2(aq) + 8H_2O(l).$$
> By itself, hydrogen peroxide decomposes slowly in accordance to the following equation:
> $$2H_2O_2(aq) \rightarrow 2H_2O(l) + O_2(g).$$
> It is found that the reaction is first-order with respect to H_2O_2. The reaction can be accelerated by using manganese(IV) oxide as the catalyst. You are supposed to plan an experiment to determine the rate of reaction for the decomposition of hydrogen peroxide with the following chemicals and apparatus:
>
> - 50 cm^3 of 0.010 mol dm^{-3} acidified $KMnO_4$;
> - 100 cm^3 of 0.025 mol dm^{-3} H_2O_2 solution;
> - Solid manganese(IV) oxide;
> - Stopwatch; and
> - Standard titration glassware in the lab.

> **Q** Where should we start thinking?

A: (i) What is the purpose of the plan?
— To determine the rate of decomposition of H_2O_2.

(ii) What do you need to know in order to determine the rate of decomposition of H_2O_2?
— We need to know how fast the H_2O_2 has reacted with respect to time.

(iii) How can you then measure the rate?
— As the hydrogen peroxide decomposes, its concentration would decrease. Hence, at specific time intervals, we can determine the concentration of hydrogen peroxide through titrating a small aliquot of the H_2O_2 solution with the acidified $KMnO_4$ solution.

(iv) So, how many data points are you going to get?
— Since we are going to plot a graph, we need at least five titration points.

(v) Do you need to dilute the hydrogen peroxide solution?
— Let us check with some pre-experimental calculations.

Pre-Experimental Calculations:

$$2MnO_4^-(aq) + 5H_2O_2(aq) + 6H^+(aq) \rightarrow 2Mn^{2+}(aq) + 5O_2(aq) + 8H_2O(l)$$

Assuming that $25\,cm^3$ of $0.025\,mol\,dm^{-3}$ H_2O_2 solution is used:

Amount of H_2O_2 present = $\dfrac{25}{1000} \times 0.025 = 6.25 \times 10^{-4}$ mol

Amount of MnO_4^- needed = $\dfrac{2}{5} \times$ Amount of H_2O_2 present = 2.5×10^{-4} mol

Volume of $0.010\,mol\,dm^{-3}$ acidified $KMnO_4$ solution used

$$= \dfrac{2.5 \times 10^{-4}}{0.010} = 0.025\,dm^3 = 25\,cm^3$$

Since the ratio of the two volumes is about 1:1, no dilution is needed.

The Procedure:

(i) Fill up the 50 cm^3 *burette* with the standard *acidified KMnO$_4$ solution*.
(ii) Pipette *10 cm^3* of the 0.025 mol dm^{-3} H$_2$O$_2$ solution into a *conical flask*.
(iii) Take the *initial burette reading*.
(iv) *Titrate* the solution *with acidified KMnO$_4$*. Swirl continuously during the addition of the titrant.
(v) Toward the *end-point*, add the KMnO$_4$ solution *dropwise* and swirl. Stop the addition of the titrant when *one drop of titrant* causes the solution in the conical flask to change from *colorless to pale pink*.
(vi) Record the *final burette reading*. *Calculate* the volume of acidified KMnO$_4$ solution needed to react with the hydrogen peroxide solution. This titer volume would indirectly give us the original concentration of the H$_2$O$_2$ solution before a substantial amount of decomposition takes place.
(vii) Use a *measuring cylinder* to measure *80 cm^3* of the 0.025 mol dm^{-3} H$_2$O$_2$ solution into a *clean and dry conical flask*.
(viii) Add *some solid manganese(IV) oxide* into the conical flask. *Start the stopwatch. Swirl gently.*
(ix) At the *5-, 10-, 15-, 20-, and 25-min intervals, pipette 10 cm^3-aliquots* of the solution and quickly *add 20 cm^3 of cold water*.
(x) *Repeat titration steps (iii) to (vi)*.

Time/min	0	5	10	15	20	25
Final burette reading/cm^3						
Initial burette reading/cm^3						
Titer volume/cm^3						

Q What is an aliquot?

A: Aliquot refers to a portion of a larger volume of solution.

> **Q** Why do you pipette 10 cm³ of the aliquot and not 25 cm³?

A: Using more solution required more time for the completion of the titration before the next one. Hence, 10 cm³ is a good choice.

Treatment of Results:

$$2MnO_4^-(aq) + 5H_2O_2(aq) + 6H^+(aq) \rightarrow 2Mn^{2+}(aq) + 5O_2(aq) + 8H_2O(l)$$

Amount of MnO_4^- used $= [MnO_4^-] \times V_{titer}$, where V_{titer} is the titer volume

Amount of $H_2O_2 = \frac{5}{2} \times$ Amount of MnO_4^- used $= \frac{5}{2} \times [MnO_4^-] \times V_{titer}$

Concentration of $H_2O_2 = \dfrac{\text{No. of moles of hydrogen peroxide}}{\text{Volume of solution pipetted}}$, where the volume of solution pipetted is always kept at 10 cm³; hence, $[H_2O_2] \propto V_{titer}$.

Therefore, plotting a graph of V_{titer} against time is equivalent to plotting $[H_2O_2]$ against time. From the graph, we would be able to determine the order of reaction for the decomposition of hydrogen peroxide.

> **Q** So, from the graph of V_{titer} against time, does the gradient give us the rate of reaction?

A: The rate of reaction has a unit of mol dm⁻³ time⁻¹. The gradient of the V_{titer} against time graph has a unit of volume per unit time. This is proportional to the rate but *not* equal to the rate. So, you need to do some conversion here.

> **Q** Can we use the same plan to determine the rate constant, k, for the rate equation: rate = k[H₂O₂]?

A: Yes, certainly! Just simply follow the same procedure to obtain a V_{titer} against time graph. Then at the $t = 0$ point, draw a tangent to determine the rate of volume change of the acidified KMnO₄ used. In order to calculate the actual instantaneous rate, you need to convert the rate of volume change to the rate of change of the $[H_2O_2]$. Now, with the instantaneous rate and the initial concentration of H_2O_2, you can substitute them into the equation, rate $= k[H_2O_2]$, to find the rate constant, k.

 Q Is there any other way that we can determine the rate constant, k, from the graph?

A: For this case here, since you have already known that it is a first-order reaction, you can determine the half-life for the reaction from the graph. Then, use the equation, $t_{1/2} = \dfrac{\ln 2}{k}$, to calculate the rate constant, k.

Note: You can also use the same planning outline for:

(i) Determine the rate of reaction of chloroethane with sodium hydroxide.

The Task:

Chloroethane, CH_3CH_2Cl, undergoes nucleophilic substitution reaction with hydroxide ion:

$$CH_3CH_2Cl + OH^- \rightarrow CH_3CH_2OH + Cl^-.$$

There are two possible rate equations: (1) rate = $k[CH_3CH_2Cl]$; or
(2) rate = $k[CH_3CH_2Cl][OH^-]$.

Design an experiment that will enable you to determine the order of reaction with respect to OH^-. You are provided with the following chemicals and apparatus:

- 0.150 mol dm^{-3} of aqueous CH_3CH_2Cl;
- 0.030 mol dm^{-3} of aqueous NaOH;
- 0.030 mol dm^{-3} of aqueous HCl;
- Methyl orange indicator;
- Stopwatch; and
- Standard titration glassware in the lab.

(Note: Similar to the titrimetric determination of the rate of decomposition of hydrogen peroxide, here, small aliquots of solution with OH^- present is withdrawn and titrated against the HCl solution. Plot the volume of HCl used versus time. If the shape of the graph is a straight line, the reaction is zero order with respect to $[OH^-]$ (rate = $k[CH_3CH_2Cl]$). If it is a curve with a constant half-life, then the reaction is first-order with respect to $[OH^-]$(rate = $k[CH_3CH_2Cl][OH^-]$).

5.4 Determine the Order of Reaction Using the Initial Rate Method

The Task:

Thiosulfate undergoes decomposition to form sulfur and sulfur dioxide in an acidic medium:

$$S_2O_3^{2-}(aq) + 2H^+(aq) \rightarrow S(s) + SO_2(g) + H_2O(l).$$

You are to design an experiment to determine the order of reaction with respect to $S_2O_3^{2-}$ and H^+ using the initial rate method by making use of the following chemicals and apparatus:

- 2.00 mol dm^{-3} HCl;
- 0.20 mol dm^{-3} Na$_2$S$_2$O$_3$;
- Stopwatch;
- Paper with printed text;
- 50 cm^3 measuring cylinders;
- 100 cm^3 beaker; and
- Standard glassware in the lab.

Q Where should we start thinking?

A: (i) What is the purpose of the plan?
— To determine the order of reaction with respect to $S_2O_3^{2-}$ and H^+ using the initial rate method.

(ii) What do you need to know in order to determine the order of reaction with respect to $S_2O_3^{2-}$ and H^+ using initial rate method?

— We need to vary the concentration of $S_2O_3^{2-}$ while keeping the concentration of H^+ constant. Then, we would need to measure how the rate is affected by the change in the concentration of the $S_2O_3^{2-}$. Likewise, repeat the whole experiment by changing the concentration of H^+ instead while keeping the concentration of $S_2O_3^{2-}$ constant.

(iii) How do you then measure the rate?

— As the reaction proceeds, solid sulfur is produced. Thus, we can measure the time taken for a "fixed" amount of sulfur to cover a piece of paper with printed text which has been placed underneath the beaker.

The Procedure:

(i) Use a *measuring cylinder* to measure *10 cm³ of HCl* into a *beaker*.
(ii) Use another *measuring cylinder* to measure *20 cm³ of water* into the beaker.
(iii) Use a different *measuring cylinder* to measure *10 cm³ of $Na_2S_2O_3$* into the beaker.
(iv) *Start the stopwatch* immediately. *Swirl* the contents and place the beaker over the printed material.
(v) *Stop the stopwatch the moment a layer of solid sulfur masks the printed material.*
(vi) *Record the time taken, t,* in the table below.
(vii) Throw the contents away, wash the beaker, and get ready to repeat the next set of experiment according to the table below.

Expt. No.	Volume of HCl/cm³	Volume of $Na_2S_2O_3$/cm³	Volume of H_2O/cm³	Total Volume/cm³	Time, t/s	$1/t$ / s^{-1}
1	10	10	20	40		
2	20	10	10	40		
3	10	20	10	40		

176 *Understanding Experimental Planning for Advanced Level Chemistry*

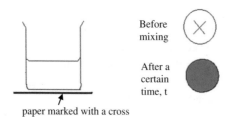

paper marked with a cross

Q Can we swirl the beaker throughout the whole experiment?

A: No! If you swirl the beaker throughout, you will not get a layer of sulfur to mask the printed material.

Q So, for each of the experiment, the layer of sulfur needed to mask the printed material is of about the same thickness?

A: You are right. The layer of sulfur is similar to a 100-m race. If all the runners travel the same distance, then the speed $\propto \dfrac{1}{\text{time taken to cover the 100 m distance}}$. Similarly, if Δc is kept constant (it is the layer of sulfur here), then the rate $\propto \dfrac{1}{\Delta t}$.

Q Why must you add some water for each experiment?

A: If you mix two solutions together, the final concentration is not the same as the initial concentration before mixing because of dilution. If you refer to Experiments 1 and 2, although the volumes of the $Na_2S_2O_3$ solution used are the same, the new concentrations of the $Na_2S_2O_3$ are different because the total volumes of the two mixtures are not the same. Hence, if we add some water to maintain the same total volume, we make sure that the new concentration of each reactant is actually directly proportional to the volume that you have measured. Meaning? As the volume of HCl solution measured for Experiment 2 is double that of Experiment 1, the new concentration of HCl after mixing for Experiment 2 is also double that for Experiment 1, i.e., concentration $\propto V_{used}$. The following calculations would convince you that the relationship holds:

Given that the initial concentration of $S_2O_3^{2-}$ is 0.20 mol dm^{-3} and the initial concentration of H$^+$ is 2.00 mol dm^{-3}:

$$[S_2O_3^{2-}]_{after\ mixing} = \frac{0.20 \times V_{thiosulfate}}{V_{total}}$$

Hence, $[S_2O_3^{2-}]_{after\ mixing} \propto V_{thiosulfate}$

$$[H^+]_{after\ mixing} = \frac{2.00 \times V_{acid}}{V_{total}}$$

Hence, $[H^+]_{after\ mixing} \propto V_{acid}$.

Treatment of Results:

Method 1: Using $\frac{1}{t}$

Comparing Experiments 1 and 2,

When V is doubled, $\frac{1}{t}$ does not change.

$\Rightarrow \frac{1}{t}$ does not change with change in V;

$\Rightarrow$ Rate does not change with change in [HCl]; and

$\Rightarrow$ Order of reaction with respect to HCl = 0.

OR

Comparing Experiments 1 and 2,

When V is doubled, $\frac{1}{t}$ is doubled

$\Rightarrow \frac{1}{t} \propto V$;

$\Rightarrow$ Rate $\propto$ [HCl]; and

$\Rightarrow$ Order of reaction with respect to HCl = 1.

OR

Comparing Experiments 1 and 2,

When V is doubled, $\frac{1}{t}$ is increased 4 times

$\Rightarrow \frac{1}{t} \propto V^2;$

$\Rightarrow$ Rate $\propto [HCl]^2$; and

O order of reaction with respect to HCl = 2.

Method 2: Using the '$V^n t$ = constant' concept

We can also use $V^n t$ = constant to determine the order of the reaction. But how does it work?

Knowing that

(1) rate $\propto \frac{1}{t} \Rightarrow$ rate $= \frac{\alpha}{t}$ S, where α is a proportionality constant and

(2) $[A] \propto V_{used} \Rightarrow [A] = \beta V_{used}$, where β is a proportionality constant,

Substituting (1) and (2) into rate = $k[A]^n \Rightarrow \frac{\alpha}{t} = k(\beta V_{used})^n$; and

$$\Rightarrow V_{used}^n \cdot t = \frac{\alpha}{\beta \cdot k} = \text{constant}.$$

Thus, if

(1) t = constant for different V values $\Rightarrow n = 0 \Rightarrow$ zero-order reaction;

(2) Vt = constant for different V values $\Rightarrow n = 1 \Rightarrow$ first-order reaction;

(3) $V^2 t$ = constant for different V values $\Rightarrow n = 2 \Rightarrow$ second-order reaction; and

(4) $V^n t$ = constant for different V values $\Rightarrow n = n \Rightarrow n^{th}$ order reaction.

Expt. No.	Volume of HCl/cm³	Volume of Na₂S₂O₃/cm³	Volume of H₂O/cm³	Total Volume/cm³	Time, t/s	1/t s⁻¹	Vt/ cm³ s	V²t/ cm⁶ s
1	10	10	20	40				
2	20	10	10	40				
3	10	20	10	40				

Method 3: Using the graphical approach

In order to use this approach, there must at least be five data points. But how does it work?

If the rate equation is, rate = k[A]^n $\Rightarrow \frac{1}{t} \propto V^n$.

Thus, if (1) $n = 0$, plotting a graph of $\frac{1}{t}$ versus V would give us a horizontal straight line;

(2) $n = 1$, plotting a graph of $\frac{1}{t}$ versus V would give us a linear straight line;

(3) $n = 2$, plotting a graph of $\frac{1}{t}$ versus V^2 would give us a linear straight line; and

(4) $n = n$, plotting a graph of $\frac{1}{t}$ versus V^n would give us a linear straight line.

5.5 Determine the Order of Reaction of $Na_2S_2O_8$ and NaI Using the Initial Rate Method

The Task:

Peroxodisulfate ions, $S_2O_8^{2-}$, react with iodide, I^-, as follows:

$$S_2O_8^{2-}(aq) + 2I^-(aq) \rightarrow I_2(aq) + 2SO_4^{2-}(aq).$$

The rate of the reaction can be studied by adding a fixed amount of $Na_2S_2O_3$ into the system and measuring the time taken for the I_2 produced to react with the $S_2O_3^{2-}$:

$$I_2(aq) + 2S_2O_3^{2-}(aq) \rightarrow 2I^-(aq) + S_4O_6^{2-}(aq).$$

You are to design an experiment to determine the order of reaction with respect to $S_2O_8^{2-}$ and I^- using the initial rate method by making use of the following chemicals and apparatus:

- Aqueous NaI;
- Aqueous $Na_2S_2O_3$;

- Aqueous $Na_2S_2O_8$;
- Starch solution;
- Stopwatch;
- 50 cm³ measuring cylinders;
- 100 cm³ beaker; and
- Standard glassware in the lab.

Q Where should we start thinking?

A: (i) What is the purpose of the plan?
— To determine the order of reaction with respect to $S_2O_8^{2-}$ and I^- using the initial rate method.

(ii) What do you need to know in order to determine the order of reaction with respect to $S_2O_8^{2-}$ and I^- using the initial rate method?
— We need to vary the concentration of $S_2O_8^{2-}$ while keeping the concentration of I^- constant. Then, we need to measure how the rate is affected by the change in the concentration of the $S_2O_8^{2-}$. Likewise, repeat the whole experiment by changing the concentration of I^- instead while keeping the concentration of the $S_2O_8^{2-}$ constant.

(iii) How can you then measure the rate?
— As the following reaction proceeds, I_2 is produced. Thus, we can measure the time taken for a "fixed" amount of $S_2O_3^{2-}$ to react with a fixed amount of I_2 produced:

$$I_2(aq) + 2S_2O_3^{2-}(aq) \rightarrow 2I^-(aq) + S_4O_6^{2-}(aq).$$

The Procedure:

(i) Use different *measuring cylinders* to measure 5 cm³ of KI, 5 cm³ of $Na_2S_2O_3$, and 40 cm³ of water into the *beaker*.
(ii) Add two drops of *starch solution* into the beaker.

(iii) Use a *measuring cylinder* to measure 10 cm³ of $Na_2S_2O_8$ solution and add it into the beaker.
(iv) *Start the stopwatch* immediately. *Swirl* the content and place the beaker over a *white tile*.
(v) *Stop the stopwatch the moment a blue-black coloration appears.*
(vi) *Record the time taken, t, in the table below.*
(vii) Throw the contents away, wash the beaker, and get ready to repeat the next set of experiment according to the table below.

Expt. No.	Volume of NaI/cm³	Volume of $Na_2S_2O_3$/cm³	Volume of H_2O/cm³	Volume of $Na_2S_2O_8$,V/cm³	Total Volume/cm³	Time, t/s	1/t s⁻¹
1	5	5	40	10	60		
2	5	5	30	20	60		
3	5	5	20	30	60		
4	5	5	10	40	60		
5	5	5	0	50	60		

(viii) Repeat the experiment by using different volumes of KI as follows:

Expt. No.	Volume of NaI, V/cm³	Volume of $Na_2S_2O_3$/cm³	Volume of H_2O/cm³	Volume of $Na_2S_2O_8$/cm³	Total Volume/cm³	Time, t/s	1/t s⁻¹
1	10	5	40	5	60		
2	20	5	30	5	60		
3	30	5	20	5	60		
4	40	5	10	5	60		
5	50	5	0	5	60		

Treatment of Results:

Since there are five data points for each reactant, you can plot graphs to determine the order of the reaction with respect to each of the reactants (refer to Section 5.3). Check to see whether the profile of the graph resembles one of the following:

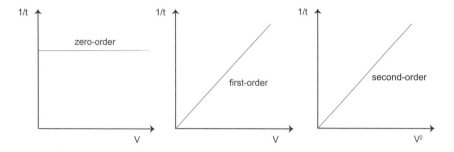

Note: You can also use the same planning outline for:

(i) Determine the order of reaction of H_2O_2 and NaI using the initial rate method.

The Task:

Hydrogen peroxide, H_2O_2, reacts with iodide, I^- as follows:

$$H_2O_2(aq) + 2I^-(aq) + 2H^+(aq) \rightarrow 2H_2O(l) + I_2(aq).$$

The rate of the reaction can be studied by adding a fixed amount of $Na_2S_2O_3$ into the system and then measuring the time taken for a small amount of I_2 produced to react with the $S_2O_3^{2-}$:

$$I_2(aq) + 2S_2O_3^{2-}(aq) \rightarrow 2I^-(aq) + S_4O_6^{2-}(aq)$$

You are to design an experiment to determine the order of reaction with respect to $S_2O_2^{2-}$ and I^- using the initial rate method by making use of the following chemicals and apparatus:

- Aqueous NaI;
- Aqueous $Na_2S_2O_3$;
- Aqueous H_2O_2;
- Aqueous HCl;
- Starch solution;
- Stopwatch;

- 50 cm³ measuring cylinders;
- 100 cm³ beaker; and
- Standard glassware in the lab.

(Note: For this reaction, we must ensure that the HCl does not come in contact with the $Na_2S_2O_3$ before the H_2O_2 reacts with the NaI. This is because the H^+ would react with the $S_2O_3^{2-}$ as follows:

$$S_2O_3^{2-}(aq) + 2H^+(aq) \rightarrow S(s) + SO_2(g) + H_2O(l).$$

Hence, it is suggested that the HCl solution and the NaI solution are mixed as one solution first before adding them into the system to start the reaction.)

(ii) Determine the order of reaction of iodide and iodate ions using the initial rate method.

The Task:

In the presence of hydrogen ions, iodate(V) ions, IO_3^-, oxidize iodide ions to iodine:

$$IO_3^-(aq) + 5I^-(aq) + 6H^+(aq) \rightarrow 3I_2(aq) + 3H_2O(l).$$

The rate of the reaction can be studied by adding a fixed amount of $Na_2S_2O_3$ into the system and then measuring the time taken for a small amount of I_2 produced to react with the $S_2O_3^{2-}$:

$$I_2(aq) + 2S_2O_3^{2-}(aq) \rightarrow 2I^-(aq) + S_4O_6^{2-}(aq).$$

You are to design an experiment to determine the order of reaction with respect to I^- and IO_3^- using the initial rate method by making use of the following chemicals and apparatus:

- Aqueous NaI;
- Aqueous $Na_2S_2O_3$;
- Aqueous $NaIO_3$;
- Aqueous H_2SO_4;

- Starch solution;
- Stopwatch;
- 50 cm³ measuring cylinders;
- 100 cm³ beaker; and
- Standard glassware in the lab.

(Note: For this reaction, we must ensure that the H_2SO_4 does not come in contact with the $Na_2S_2O_3$ before the $NaIO_3$ reacts with the NaI. This is because the H^+ would react with the $S_2O_3^{2-}$ as follows:

$$S_2O_3^{2-}(aq) + 2H^+(aq) \rightarrow S(s) + SO_2(g) + H_2O(l).$$

Hence, it is suggested that the H_2SO_4 solution and the NaI solution are mixed as one solution first before adding them into the system to start the reaction.)

5.6 Determine the Activation Energy, E_a, of the Reaction Between $Na_2S_2O_8$ and NaI using the Initial Rate Method

The Task:

The activation energy, E_a, of a reaction can be calculated from the following equation:

$$\ln k = \ln A - \frac{E_a}{RT},$$

where k is the rate constant, A is the Arrhenius constant, R is the molar gas constant, and T is the temperature in Kelvin.

Peroxodisulfate ions, $S_2O_8^{2-}$, react with iodide, I^- as follows:

$$S_2O_8^{2-}(aq) + 2I^-(aq) \rightarrow I_2(aq) + 2SO_4^{2-}(aq).$$

The rate of the reaction can be studied by adding a fixed amount of $Na_2S_2O_3$ into the system and then measuring the time taken for a small amount of I_2 produced to react with the $S_2O_3^{2-}$:

$$I_2(aq) + 2S_2O_3^{2-}(aq) \rightarrow 2I^-(aq) + S_4O_6^{2-}(aq).$$

You are to design an experiment to determine the E_a of the reaction between $S_2O_8^{2-}$(aq) and I^-(aq), using the initial rate method by making use of the following chemicals and apparatus:

- Aqueous NaI;
- Aqueous $Na_2S_2O_3$;
- Aqueous $Na_2S_2O_8$;
- Starch solution;
- Stopwatch;
- 50 cm^3 measuring cylinders;
- 100 cm^3 beaker; and
- Standard glassware in the lab.

Q Where should we start thinking?

A: (i) What is the purpose of the plan?
— To determine the E_a of the reaction between $S_2O_8^{2-}$ and I^- using the initial rate method.

(ii) What do you need to know in order to determine the E_a of the reaction?
— From the equation, $\ln k = \ln A - \dfrac{E_a}{RT}$; we need to find the values of the rate constant for different temperature values. Then, plot a graph of $\ln k$ versus $\dfrac{1}{T}$. The gradient of the graph is equal to $\dfrac{E_a}{R}$.

(iii) How can you determine the value of the rate constant at a particular temperature?
— The rate constant comes from the rate equation, rate $= k[S_2O_8^{2-}]^n[I^-]^m$, thus if we know the rate, the concentration of the reactants, and the orders of reaction, we would be able to calculate the rate constant.

(iv) But do you know the order of the reaction?
— Well, rate $= k[S_2O_8^{2-}]^m[I^-]^n$; if we perform the experiments at different temperature values but keep the concentration of the reactants the same for each of the experiments, then rate $\propto k$.

(v) How can you then measure the rate?
— As the following reaction proceeds, I_2 is produced. Thus, we can measure the time taken for a known amount of $S_2O_3^{2-}$ to react with a fixed amount of I_2 produced.

$$S_2O_8^{2-}(aq) + 2I^-(aq) \rightarrow I_2(aq) + 2SO_4^{2-}(aq).$$

The Procedure:

(i) Use different *measuring cylinders* to measure 5 cm³ of KI, 5 cm³ of $Na_2S_2O_3$, and 40 cm³ of water into the *beaker*.
(ii) Add two drops of *starch solution* into the beaker in a *thermostated water bath* at 298 K.
(iii) Use a *measuring cylinder* to measure 10 cm³ of $Na_2S_2O_8$ solution and let it *incubate in the thermostated water bath* at 298 K.
(iv) Add the $Na_2S_2O_8$ solution into the beaker. *Start the stopwatch* immediately. *Swirl* the contents in the beaker. *Ensure that the beaker is always in the water bath.*
(v) *Stop the stopwatch the moment a blue-black coloration appears.*
(vi) Record the time taken, t, in the table below.
(vii) Throw the contents away, wash the beaker, and get ready to repeat the next set of experiment according to the table below.

Expt. No.	Volume of NaI/cm³	Volume of $Na_2S_2O_3$/cm³	Volume of H_2O/cm³	Volume of $Na_2S_2O_8$,V/cm³	T/K	1/T /K^{-1}	Time, t/s	1/t s^{-1}
1	5	5	40	10	298			
2	5	5	40	10	308			
3	5	5	40	10	318			
4	5	5	40	10	328			
5	5	5	40	10	338			

Treatment of Results:

Since the concentrations of the reactants are constant, rate $\propto$ k.
But rate $\propto \frac{1}{t}$ as [$Na_2S_2O_3$] is constant for all experiments. Hence, k $\propto \frac{1}{t}$.

Plot a graph of $\frac{1}{t}$ versus $\frac{1}{T}$ and use the gradient to find E_a.

Note: You can also use the same planning outline for:
Determine the molar gas constant, R.

The Task:

The molar gas constant, R, of a reaction can be calculated from the following equation:

$$\ln k = \ln A - \frac{E_a}{RT}$$

where k is the rate constant, A is the Arrhenius constant, E_a is the activation energy of the reaction (51.8 kJ mol^{-1}), and T is the temperature in Kelvin.

Peroxodisulfate ions, $S_2O_8^{2-}$, react with iodide, I^- as follows:

$$S_2O_8^{2-}(aq) + 2I^-(aq) \rightarrow I_2(aq) + 2SO_4^{2-}(aq).$$

The rate of the reaction can be studied by adding a fixed amount of $Na_2S_2O_3$ into the system and then measuring the time taken for a small amount of I_2 produced to react with the $S_2O_3^{2-}$:

$$I_2(aq) + 2S_2O_3^{2-}(aq) \rightarrow 2I^-(aq) + S_4O_6^{2-}(aq).$$

You are to design an experiment to determine the value of R through the reaction between $S_2O_8^{2-}(aq)$ and $I^-(aq)$, using the initial rate method by making use of the following chemicals and apparatus:

- Aqueous NaI;
- Aqueous $Na_2S_2O_3$;
- Aqueous $Na_2S_2O_8$;
- Starch solution;
- Stopwatch;
- 50 cm^3 measuring cylinders;
- 100 cm^3 beaker; and
- Standard glassware in the lab.

(Note: Everything is exactly the same! We just tweaked the objective of the question.)

5.7 Determine How the Concentration of an Acid Affects the Rate of Reaction of Magnesium

The Task:

You are to investigate the effect of varying the concentration of an acid on the rate of the reaction between magnesium and hydrochloric acid at room temperature using the initial rate method, in accordance to the following balanced equation:

$$HCl(aq) + Mg(s) \rightarrow MgCl_2(aq) + H_2(g).$$

Plan an experiment, which involves graph plotting, to determine the order of reaction with respect to HCl. You are provided with the following chemicals and apparatus:

- 2.00 mol dm^{-3} HCl solution;
- 25 cm long of magnesium ribbon of mass 2.5 g;
- Burettes;
- Ruler;
- Stopwatch;
- 100 cm^3 beaker; and
- Standard glassware in the lab.

Q Where should we start thinking?

A: (i) What is the purpose of the plan?
— To determine the order of reaction with respect to HCl using the initial rate method.

(ii) What do you need to know in order to determine the order of reaction with respect to HCl using the initial rate method?
— We need to vary the concentration of HCl while keeping the amount of magnesium metal used constant. Then, we need to measure how the rate is affected by the change in the concentration of the HCl solution.

(iii) How can you then determine how much of the magnesium metal to use?
- Since we need to plot a graph, we need to perform at least five sets of experiments. The magnesium ribbon can be cut into 10 pieces, each about 2.5 cm with a mass of 0.25 g. Thus, we would assume that the mass of magnesium used is 0.25 g.

(iv) How would you determine the amount of acid needed?
- Knowing the mass of magnesium needed and making use of the following equation, we would be able to know the number of moles of HCl needed, and hence, the volume of HCl solution:

$$2HCl(aq) + Mg(s) \rightarrow MgCl_2(aq) + H_2(g).$$

(v) Since you need different concentrations of HCl, how would you prepare the diluted HCl solutions?
- We can use a burette to add a specific volume of deionized water into a specific volume of HCl solution that has been taken from the stock solution.

Pre-Experimental Calculations:

Assuming that a magnesium ribbon with a length of 2.5 cm and a mass of 0.25 g is used:

$$2HCl(aq) + Mg(s) \rightarrow MgCl_2(aq) + H_2(g).$$

Amount of magnesium in 0.25 g = $\frac{0.25}{24.3}$ = 1.03×10^{-2} mol

Amount of HCl needed = 2 × Amount of magnesium in 0.25 g = 2.06×10^{-2} mol

Volume of 2.00 mol dm^{-3} HCl solution = $\frac{2.06 \times 10^{-2}}{2.00}$ = 1.03×10^{-2} dm^3
= 10.3 cm^3

Thus, the volume of 2.00 mol dm^{-3} HCl solution used must be greater than 10.3 cm^3.

To change the concentration of the HCl solution, we need to ensure that the total volume of the solution is constant for the following five sets of experiments:

Expt. No.	1	2	3	4	5
Volume of HCl/cm^3	20.00	30.00	35.00	40.00	50.00
Volume of Water/cm^3	30.00	20.00	15.00	10.00	0.00
Total Volume/cm^3	50.00	50.00	50.00	50.00	50.00

The minimum volume of HCl solution is 20 cm^3, which ensures that the H$^+$ ions are in excess.

The Procedure:

(i) Take a piece of *sandpaper* to *gently polish the surfaces of the magnesium ribbon*.

(ii) *Cut the ribbon* into 10 pieces, each *with a length* of 2.5 cm.

(iii) Use a 50 cm^3 *burette* to measure 20 cm^3 of 2.00 mol dm^{-3} *HCl solution* into a *clean and dry beaker*.

(iv) Use another 50 cm^3 *burette* to introduce 30 cm^3 of *deionized water* into the same beaker.

(v) *Stir* the solution to mix well. Place a piece of the *magnesium ribbon* into the acid solution and immediately *start the stopwatch*.

(vi) *Gently stir the solution continuously* until the magnesium disappears. Immediately *stop the stopwatch*.

(vii) *Record the time taken* for the piece of magnesium ribbon to dissolve in the acid.

(viii) *Repeat steps (iii) to (vii)* with an acid of different concentration as shown in the table above.

Q Why do you need to polish the magnesium ribbon with the sandpaper?

A: The surface of the magnesium ribbon is usually coated with some magnesium oxide. The sandpapering is to get rid of the layer of oxide.

> **Q** Why do you need to stir the solution continuously?

A: As magnesium reacts with H$^+$, the H$^+$ ions surrounding the magnesium ribbon depletes. Diffusion needs to take place in order for other H$^+$ ions to come into contact with the magnesium again. If the concentration of the acid used is different, the diffusion rate may affect the actual rate of reaction. So, stirring is necessary to ensure that there is a "constant" concentration of H$^+$ ions surrounding the magnesium ribbon.

> **Q** Other than waiting for the magnesium ribbon to dissolve in the acid, can we use a gravimetric or gas collection method to monitor the rate of reaction?

A: Certainly you can. You can place the set-up on a weighing balance and measure the time taken for the mass to reach a constant value. The hydrogen gas will escape; hence, the mass of the set-up would decrease. Or, you can measure the time taken to collect a fixed volume of hydrogen gas, using the gas collection method.

5.8 Determine the Order of Reaction of Iodination of Propanone Using Colorimetry

The Task:

In the presence of an acid catalyst, HCl, propanone is iodinated as follows:

$$CH_3COCH_3(aq) + I_2(aq) \rightarrow CH_3COCH_3I(aq) + HI(aq).$$

As the reaction proceeds, the brown color of iodine gradually turns colorless. This is because as the iodine is consumed in the reaction, its concentration decreases. Hence, the rate of the reaction can be monitored by measuring the absorbance value (A) using a colorimeter. You are to design an experiment to determine the order of reaction with respect to

CH$_3$COCH$_3$ and I$_2$, using the initial rate method by making use of the following chemicals and apparatus:

- 2.00 mol dm^{-3} propanone;
- 0.020 mol dm^{-3} aqueous iodine in KI;
- 1.00 mol dm^{-3} aqueous HCl;
- Colorimeter;
- Measuring cylinders;
- 100 cm^3 beaker; and
- Standard glassware in the lab.

Q Why is there KI in the aqueous iodine solution?

A: Being a non-polar molecule, the solubility of iodine in water is very low. But the solubility can be enhanced due to the formation of the I$_3^-$ species:

$$I_2(aq) + I^-(aq) \rightleftharpoons I_3^-(aq) \text{ (brown-colored complex)}.$$

Q Where should we start thinking?

A: (i) What is the purpose of the plan?
— To determine the order of reaction with respect to CH$_3$COCH$_3$ and I$_2$ using the initial rate method.

(ii) What do you need to know in order to determine the order of reaction with respect to CH$_3$COCH$_3$ and I$_2$ using the initial rate method?
— We need to vary the concentration of CH$_3$COCH$_3$ while keeping the concentration of I$_2$ constant. Then, we need to measure how the rate is affected by the change in the concentration of the CH$_3$COCH$_3$. Likewise, repeat the whole experiment by changing the concentration of I$_2$ instead while keeping the concentration of the CH$_3$COCH$_3$ constant.

(iii) How can you then measure the rate?
— Since the initial rate $\frac{\Delta c}{\Delta t}$, as the following reaction proceeds:

$$CH_3COCH_3(aq) + I_2(aq) \rightarrow CH_3COCH_3I(aq) + HI(aq),$$

I_2 is consumed. Hence, we can measure the amount of iodine consumed ($\Delta c \propto \Delta A$) in a fixed amount of time, i.e., Δt is fixed.

The Procedure:

(i) Use different *measuring cylinders* to introduce 10 cm³ of propanone, 5 cm³ of acid, 20 cm³ of water, and 5 cm³ of iodine together in a *colorimeter*.
(ii) *Start the stopwatch* and *mix* the solution well.
(iii) *Measure the change of absorbance* (ΔA) *reading for a duration of 30 s*.
(iv) *Repeat the experiments* using different volumes of the reactants.

Expt No.	Volume of Propanone/cm³	Volume of I_2(aq)/cm³	Volume of H_2O/cm³	Volume of HCl/cm³	Total Volume/ cm³	ΔA
1	10	5	20	5	40	
2	20	5	10	5	40	
3	10	10	15	5	40	

Treatment of Results:

(i) To determine the order of reaction with respect to propanone:

As rate $\propto \Delta c$, where $\Delta c \propto \Delta A$, the order of the reaction can be deduced by comparing the ΔA values for Experiments 1 and 2.

(ii) To determine the order of reaction with respect to iodine:

Since Δt is kept constant for Experiments 2 and 3, do not forget that the [I_2] is doubled. Hence, rate $\propto \frac{\Delta c}{\Delta t} \propto \frac{\Delta A}{\Delta t}$, and NOT rate $\propto \Delta A$.

> **Q** Can we use the continuous method to measure the rate using colorimetry?

A: Yes. Simply measure the absorbance (A) continuously with respect to time. Then, plot an absorbance versus time graph.

> **Q** As iodine can react with thiosulfate, can we use a titrimetric approach for the continuous method?

A: Oh, yes. But you need something to stop the reaction from proceeding during the titration. One way is to add sodium hydrogencarbonate ($NaHCO_3$) which would react with all the acid. The procedure would be as follows:

(i) Fill up the 50 cm^3 *burette* with the standard *sodium thiosulfate solution*.
(ii) Use different 50 cm^3 *burettes* to introduce 30 cm^3 of iodine solution and 10 cm^3 of acid into a *clean and dry conical flask*. Add 20 cm^3 of deionized water. *Swirl* and mix the contents well.
(iii) *Pipette* 10 cm^3 of the mixture and *titrate with standard sodium thiosulfate* solution. When the solution *turns pale yellow*, add two drops of starch indicator. Titrate until the *blue-black coloration has decolorized*. The titer value would give us the amount of iodine at $t = 0$.
(iv) Use different 50 cm^3 *burettes* to introduce 30 cm^3 of iodine solution and 10 cm^3 of acid into a *clean and dry conical flask*.
(v) Add 20 cm^3 of *propanone. Start the stopwatch* immediately. *Swirl* and mix the contents well.
(vi) At the 5-, 10-, 15-, 20- and 25-min intervals, *pipette 10 cm^3 aliquots of the solution* and quickly add 20 cm^3 of *aqueous $NaHCO_3$ solution*.
(vii) *Titrate the leftover iodine using the standard sodium thiosulfate*.

5.9 Safety Precautions for Kinetics Experiments

You may be asked to quote some safety precautions while performing a kinetics experiment. Depending on the type of kinetics experiment that you are performing, the following examples may be useful for you to take note:

— Always wear gloves, a lab coat, and safety goggles while doing the experiment. For example, solid sodium hydroxide, acids (H_2SO_4) or bases (NaOH), methanol, etc. that you used may be corrosive in

nature. So, there should be minimal direct contact of the skin with these chemicals.
— If there are any toxic gases or fumes evolved, conduct the experiment in a fumehood as such fumes may cause respiratory problems.
— If there are flammable liquids used, such as alcohol, ensure that there is no naked flame around.
— Any organic chemicals should be properly covered or capped when not in use and used organic waste must be properly disposed.

5.10 Minimizing Experimental Errors or Increasing Reliability

You may be asked to quote some ways to improve experimental reliability while performing a kinetics experiment. Depending on the type of kinetics experiment that you are performing, the following examples may be useful for you to take note:

— Some of the compounds are hydroscopic in nature. Weigh the compound quickly or cover it after weighing to avoid the absorption of moisture from the air.
— The solution to initiate the reaction must be poured into the reactor using a measuring cylinder as we need to let all the reactants come in contact in the shortest time.
— Human reaction lag time during the starting and stopping of the stopwatch is an issue in kinetics study.
— If the volume of the solution needs to be more precise, use a pipette or burette instead of a measuring cylinder as the latter is a less-precise apparatus.
— If the reaction needs to be quenched, you can add some cold water to slow down the reaction. This is useful if the reaction is being carried out at a temperature that is much higher than that of the cold water. If not, you can "kill" the reaction by adding some excess reagent which would react with one of the reactants. Then, perform a back-titration to determine the amount of the leftover reagent.
— The experiment must always be repeated to check for reliability of results. The average value should be calculated if necessary.

CHAPTER 6

PLANNING FOR ELECTROCHEMICAL EXPERIMENTS

In a redox reaction, the electrons released during an oxidation process are consumed during a reduction process. If these electrons are allowed to move through an external circuit before reduction, we have an electric current. Such a cell that converts chemical energy to electrical energy is known as an electric cell or voltaic cell.

In contrast, the process of passing an electric current to force an otherwise *non-spontaneous* redox reaction to occur is known as electrolysis, and the set-up is termed the *electrolytic cell*.

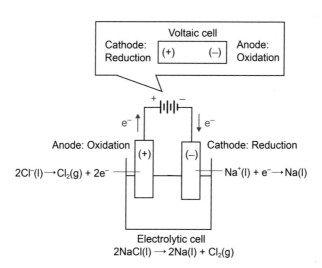

In the electrolytic cell, we have:

- The *electrolyte*, which is a compound in *solution* or a *molten* compound that is able to conduct electricity because there are *mobile* charge carriers in the form of ions. The electrolyte is decomposed in the process;
- An electric cell or power source, which is used to drive the flow of electrons in the electrolytic set-up. It acts like a "pump" to push electrons in a single direction;
- The *electrodes*, which are metal conductors or graphite, by which an electric current enters or leaves the electrolyte;
- The *cathode*, which is the *reduction* electrode and is *negatively* charged because it is *connected* to the *negative* terminal of the electric cell. Take note that the polarity or sign of the cathode for an electrolytic cell is *opposite* of that for an electric cell; and
- The *anode*, which is the *oxidation* electrode and is *positively* charged because it is *connected* to the *positive* terminal of the electric cell. Take note that the polarity or sign of the anode for an electrolytic cell is *opposite* of that for an electric cell.

During electrolysis, not all species in the system can undergo oxidation or reduction. In fact, there is selective discharge of species, which is dependent on the standard reduction potential value, E^θ (refer to *Understanding Advanced Physical Inorganic Chemistry* by J. Tan and K. S. Chan). The more positive the E^θ value,

$\Rightarrow$ the more likely the oxidized state is to undergo reduction at the cathode, but
$\Rightarrow$ the less likely the reduced form will undergo oxidation at the anode.

> **Q** Why is the more positive the E^θ value, the more likely the reduction would take place?

A: The spontaneity of a reaction is measured by the change of the Gibbs Free Energy, ΔG^θ. The more negative the value of ΔG^θ, the more spontaneous is the reaction. We can link up E^θ value with ΔG^θ as in the equation,

$\Delta G^\theta = -nFE^\theta$, where n is the number of moles of electrons transferred during the redox reaction as per balanced chemical equation and F is the Faraday's constant. From the equation, the more positive the E^θ value, the more negative is the ΔG^θ value and hence the more spontaneous is the redox reaction.

Faraday's Laws of Electrolysis:

Faraday's laws of electrolysis provide us with the quantitative relationship between electricity and chemical change.

Faraday's First Law of Electrolysis: The *amount of material* (mass and/or volume) that is being discharged at each of the electrodes depends solely on the amount of charges that has passed through the system. Thus, using the following relationship, we can perform quantitative calculations for an electrolytic reaction:

$Q = I.t$ or $Q = nF$, where Q = quantity of charge in coulombs, C;
I = current in amperes, A or C s^{-1};
t = time in seconds, s;
F = Faraday's constant (96,500 C mol^{-1}); and
n = number of moles of electrons

Faraday's Second Law of Electrolysis: The *amount of charge* that is required to discharge one mole of an element depends on the charge, z, on the ion. The amount of charge, Q, in coulombs (C) is the product of the current, I, in amperes (A or C s^{-1}) and time, t, in seconds (s), i.e.,

$$Q = I \times t.$$

1 C of charge is the amount of electric charge passed by a current of 1 A in 1 s. Therefore, 1 Faraday (F) is the amount of charge (in C mol^{-1}) carried by one mole of electrons:

F = L.e   where L = Avogadro's constant = 6.02×10^{23} mol^{-1};

e = charge on an electron = -1.60×10^{-19} C; and
F = 96,500 C mol^{-1}.

In other words, the amount of electrons = $\frac{Q}{96,500}$ mol.

Hence, the amount of charge that is required to discharge one mole of the following ions has been experimentally found to be:

1 mol of Ag^+: $1 \times 965 \times 10^4$ C since $Ag^+ + 1e^- \rightarrow Ag$;
1 mol of Cu^{2+}: $2 \times 965 \times 10^4$ C since $Cu^{2+} + 2e^- \rightarrow Cu$; and
1 mol of Al^{3+}: $3 \times 965 \times 10^4$ C since $Al^{3+} + 3e^- \rightarrow Al$.

Therefore, one mole of Ag^+, Cu^{2+}, and Al^{3+} requires 1, 2, and 3 Faradays, respectively, to discharge at the cathode.

It looks like both Faraday's first and second laws have no differences; is it true?

A: Not really. The first law is concerned with how much material you would get after a particular amount of charge has passed through the electrolytic cell. The second law is concerned with the amount of charge needed to discharge one mole of ions. That means, the first law talks about the quantity of material whereas the second law talks about the quantity of charge; there is a different focus. Hence, you may be asked to design questions to verify these two laws.

6.1 Determine the Cell Potential of a Voltaic Cell

The Task:

The standard reduction potential values of two half-cells measured with respect to the standard hydrogen electrodes are:

$$Sn^{2+}(aq) + 2e^- \rightleftharpoons Sn(s), \quad E^\theta = +0.15 \text{ V; and}$$
$$Fe^{2+}(aq) + 2e^- \rightleftharpoons Fe(s), \quad E^\theta = -0.44 \text{ V}$$

You are supposed to plan an experiment to determine the cell potential when the following redox reaction takes place under standard conditions (25°C, 1 bar):

$$Sn^{2+}(aq) + Fe(s) \rightarrow Fe^{2+}(aq) + Sn(s).$$

Draw a diagram of your set-up. You are provided with the following chemicals and apparatus:

- 100 cm³ of 1.00 mol dm⁻³ of $SnSO_4$ solution;
- 100 cm³ of 1.00 mol dm⁻³ of $FeSO_4$ solution;

- Tin metal;
- Iron metal;
- Crocodile clips;
- Salt-bridge containing concentrated KNO$_3$ solution;
- Voltmeter; and
- Standard glassware in the lab.

Q Where should we start thinking?

A: (i) What is the purpose of the plan?
— To determine the cell potential of the following redox reaction:

$$Sn^{2+}(aq) + Fe(s) \rightarrow Fe^{2+}(aq) + Sn(s).$$

(ii) What do you need to know in order to determine the cell potential?
— We need to set up the apparatus as follows:

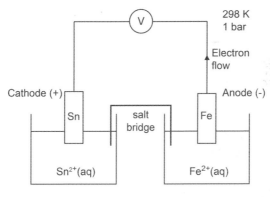

Thereafter, we need to measure the e.m.f. of the cell using the voltmeter.

The Procedure:

A typical electrochemical experiment involves the measurement of voltage (for voltaic cell) or the measurement of the amount of current that passed through a circuit in a specific amount of time for an electrolytic cell. When a student follows the experimental procedure diligently, the student should be able to obtain repeatable results. Similarly, another student doing the same experiment should obtain data that are close to what the rest have obtained. Therefore, the procedure of a typical electrochemical experiment would consist of a series of steps which inform the student: (1) the type of set-up; (2) which step should come first; (3) what apparatus should he/she use; (4) what is the quantity of substance that he/she should measure; (5) the amount of current flowing through the circuit in a specific amount of time; and (6) if needed, the reaction conditions such as temperature and pressure.

(i) Set up the apparatus as shown in the diagram above with the *switch open*.
(ii) Use a *measuring cylinder* to introduce *50 cm^3 of 1.00 mol dm^{-3} $SnSO_4$ solution* into the *beaker*.
(iii) Use another *measuring cylinder* to introduce *50 cm^3 of 1.00 mol dm^{-3} $FeSO_4$ solution* into the other beaker.
(iv) Put the *salt-bridge* in place.
(v) Measure the *temperature and pressure* using a *thermometer and barometer*.
(vi) *Turn on the switch* and take note of the *highest reading as indicated by the voltmeter*.
(vii) *Repeat the experiment* to get *reliable results*.

Q Why must we take the highest reading as indicated by the voltmeter?

A: This is because when the redox reaction starts, the concentrations of the reactants are going to decrease. The change in concentration would affect the cell potential that we are measuring. Hence, the voltage that is measured at the very beginning corresponds to the initial concentrations of the species that are known at the start.

How does the voltmeter inform us which is the anode and which is the cathode?

A: The voltmeter has a positive terminal, usually red in color, and a negative terminal, usually black in color. Let us assume that we do not know the polarity of the electrodes and you just randomly connect the terminals of the voltmeter to each of the two electrodes and you have a positive e.m.f. measured. This would mean that the electrode which is connected to the positive terminal of the voltmeter is also positive in nature, i.e., a cathode in the voltaic cell set-up, and vice versa if you obtain a negative e.m.f. measurement.

Note: You can also use the same planning outline for:

(i) Determine the e.m.f. of any redox reactions under standard conditions (1 bar, 25°C, concentration of any species is at 1 mol dm^{-3}):

$$Cu^{2+}(aq) + Zn(s) \rightarrow Zn^{2+}(aq) + Cu(s);$$
$$MnO_4^-(aq) + 8H^+(aq) + 5Fe^{2+}(aq)$$
$$\rightarrow Mn^{2+}(aq) + Fe^{3+}(aq) + 4H_2O(l);$$
$$H_2(g) + Cl_2(g) \rightarrow 2H^+(aq) + 2Cl^-(aq);$$
$$2Fe^{2+}(aq) + 2H^+(aq) \rightarrow 2Fe^{3+}(aq) + H_2(g);$$

and many more. The major difference is that when there is no metal involved in the reaction system, the inert platinum electrode is used. And when a gas species is involved, the following set-up is used:

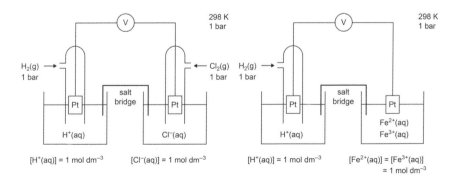

6.2 Determine the Effect of Concentration on the Cell Potential

The Task:

The standard reduction potential value of the $Cu^{2+}|Cu$ half-cell is measured with respect to the standard hydrogen electrode:

$$Cu^{2+}(aq) + 2e^- \rightleftharpoons Cu(s), \quad E^\theta = +0.34 \text{ V}.$$

The measured value is dependent on the concentration of the ion and the temperature. This means that the following redox reaction would still take place if the concentrations of the ions are different:

$$Cu^{2+}(aq) + Cu(s) \rightarrow Cu^{2+}(aq) + Cu(s).$$

You are supposed to plan an experiment to show how the cell potential changes when the concentration of the Cu^{2+} decreases. You are provided with the following chemicals and apparatus:

- 200 cm^3 of 1.00 mol dm^{-3} of $CuSO_4$ solution;
- 100 cm^3 of 0.100 mol dm^{-3} of $CuSO_4$ solution;
- 100 cm^3 of 0.010 mol dm^{-3} of $CuSO_4$ solution;
- Copper metals;
- Crocodile clips;
- Salt-bridge containing concentrated KNO_3 solution;
- Voltmeter; and
- Standard glassware in the lab.

Q Where should we start thinking?

A: (i) What is the purpose of the plan?
— To determine how the cell potential changes with a decrease in the concentration of the copper(II) ions for the following redox reaction:
$$Cu^{2+}(aq) + Cu(s) \rightarrow Cu^{2+}(aq) + Cu(s).$$

(ii) What do you need to know in order to determine the cell potential?
— Set up the apparatus as shown in Section 6.1. For one of the half-cells, fix the concentration of the $CuSO_4$ solution at 1.00 mol dm^{-3}, whereas for the other half-cell, we vary the concentration of the $CuSO_4$ solution.

The Procedure:

(i) Set up the apparatus as shown below with the *switch open*.
(ii) Use a *measuring cylinder* to introduce *50 cm^3 of 1.00 mol dm^{-3} CuSO$_4$* solution into the *beaker*.
(iii) Use another *measuring cylinder* to introduce *50 cm^3 of 0.100 mol dm^{-3} CuSO$_4$* solution into the other beaker.
(iv) Put the *salt-bridge* in place.
(v) Measure the *temperature and pressure* using a *thermometer and barometer*.
(vi) *Turn the switch on* and take note of the *highest reading as indicated by the voltmeter*.
(vii) *Repeat the experiment* using fresh CuSO$_4$ solution of concentration 1.00 mol dm^{-3}. For the other half-cell, use CuSO$_4$ solution of concentration 0.010 mol dm^{-3}.
(viii) *Repeat the entire experiment* to get *reliable results*.

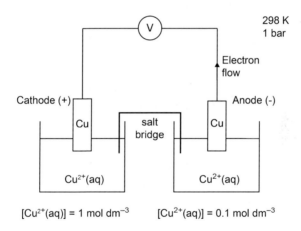

6.3 Determine Avogadro's Number Through Electrolysis

The Task:

Avogadro's number is defined as the number of particles in one mole of a substance. The currently accepted value is 6.022×10^{23} mol^{-1}. During electrolysis, the amount of material discharged at each of the electrodes depends solely on the amount of current that has passed through the system.

> Through the electrolysis of copper(II) sulfate solution using copper electrodes, plan an experiment to show how you can determine Avogadro's number. Draw the experimental set-up. You are provided with the following chemicals and apparatus:
>
> - 200 cm^3 of 1.00 mol dm^{-3} of CuSO$_4$ solution;
> - Two pieces of pure copper metals;
> - Weighing balance;
> - Crocodile clips;
> - Variable power supply;
> - Ammeter; and
> - Standard glassware in the lab.

Q Where should we start thinking?

A: (i) What is the purpose of the plan?
— To determine Avogadro's number.

(ii) What do you need to know in order to determine Avogadro's number?
— As Avogadro's number is defined as the number of particles per mole, we will need to know how many particles and the number of moles of particles formed during the electrolysis.

(iii) How do you measure the number of moles of particles that are formed during electrolysis?
— When CuSO$_4$ undergoes electrolysis using copper electrodes:

At the anode: $Cu(s) \rightarrow Cu^{2+}(aq) + 2e^-$
At the cathode: $Cu^{2+}(aq) + 2e^- \rightarrow Cu(s)$
Overall reaction: $Cu^{2+}(aq) + Cu(s) \rightarrow Cu^{2+}(aq) + Cu(s)$

If we know the mass of copper deposited at the cathode or the loss of mass of copper at the anode, using the relative atomic mass of Cu, we would be able to determine the number of moles of Cu that has been deposited at the cathode. The number of moles of electrons that has resulted in the formation of the copper metal at the cathode can now be determined.

(iv) How can you then calculate the Avogadro's number (L)?

— After knowing the number of moles of electrons that has passed through the circuit and the charge of an electron (-1.60×10^{-19} C), we would be able to calculate the amount of charge that has passed through the circuit as follows:

$$Q = \text{Number of moles of electrons} \times 1.60 \times 10^{-19}.$$

But the amount of charge that has passed through the circuit can also be calculated from the amount of current (in A or C s^{-1}) and the time taken (in second) for the flow of current as follows:

$$\text{Amount of charge, } Q = I \times t.$$

Number of electrons = L × Number of moles of electrons

Amount of charge = L × Number of moles of electrons × 1.60×10^{-19} = $I \times t$

$$\Rightarrow L = \frac{I \times t}{\text{Number of moles of electrons} \times 1.6 \times 10^{-19}} \text{ mol}^{-1}.$$

(v) How can you determine the amount of current passing through the circuit and the duration of its flow?

— Well, we can assume that the mass of copper "collected" at the cathode is 0.5 g. From here, we can calculate the number of moles of Cu atoms formed during the reduction process, and hence, the number of moles of electrons used during the reduction. Knowing the Faraday's constant, we can determine the amount of current and its duration of flow.

Pre-Experimental Calculations:

Assuming that 0.5 g of copper has deposited at the cathode:

$$Cu^{2+}(aq) + 2e^- \rightarrow Cu(s).$$

Amount of copper atoms = $\dfrac{0.5}{63.5}$ = 7.87×10^{-3} mol

Amount of electrons = 2 × Amount of copper atoms = 1.57×10^{-2} mol

Faraday's constant = 96,500 C mol^{-1}

Quantity of charge = $1.57 \times 10^{-2} \times 96{,}500 = 1.52 \times 10^{3}$ C

Assuming that 1 A of current is flowing, the duration

$$= \frac{1.52 \times 10^{3}}{1} = 1.52 \times 10^{3} \text{ s} = 25.3 \text{ min.}$$

The Procedure:

(i) Set up the apparatus as shown in the diagram below.

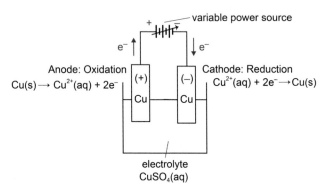

(ii) Use a *measuring cylinder* to introduce *100* cm^3 of 1.00 mol dm^{-3} CuSO$_4$ solution into the *beaker*.
(iii) *Measure the masses* of the two pieces of *pure copper metal*.
(iv) *Partially submerged* the two copper electrodes in the solution.
(v) Complete the circuit by *switching on the power supply. Set the current to 1 A. Start the stopwatch* and let the *current flow for 25 min*.
(vi) Make sure that the *amperage of the current is always maintained at 1 A. Stir the solution constantly* during the 25-min duration.
(vii) *When the time is up, switch off the power. Remove the electrode* and *rinse* it with water.
(viii) *Wipe the electrodes dry and take their masses.*
(ix) *Repeat the entire experiment* to get *reliable results*.

> **Q** Why must we constantly stir the solution during the electrolytic process?

A: When the species near the electrodes undergo discharge, a concentration gradient would be formed. The diffusion process takes time and this would affect the amperage of the current. So, constant stirring ensure a homogeneous solution throughout the electrolytic process.

Q What would happen if impure copper metal is used as the electrodes instead?

A: If that is the case, then the increase in mass for the cathode may not be equal to the decrease in mass for the anode. Why? This is because as the copper at the impure anode undergoes oxidation, the impurities may be "dislodged" at the same time. This would result in an additional decrease in mass for the anode. In addition, the impurities that undergo oxidation at the anode may not release the same amount of charges as that for copper atoms. This would overall affect the change of mass for the cathode.

Treatment of Results:

The change of mass for the cathode should be the same as that of the anode; therefore, we can take the average of the two mass changes.
Let the mass of copper deposited be m g.
Let the time be 25 min = 1,500 s and the current be 1 A.
Amount of copper atoms = $\dfrac{m}{63.5}$ = a mol
Amount of electrons = 2 × Amount of copper atoms = 2a mol
Amount of charge = $I \times t$ = 1 × 1,500 = 1,500 C
Therefore, Avogadro's constant,

$$L = \dfrac{I \times t}{\text{Number of moles of electrons} \times 1.6 \times 10^{-19}}$$

$$= \dfrac{1,500}{2a \times 1.6 \times 10^{-19}} \text{ mol}^{-1}.$$

Q Is there a graphical way to determine Avogadro's constant using the above plan?

A: Oh, yes. Refer to the following:

Amount of charge =

$L \times$ Number of moles of electrons $\times 1.60 \times 10^{-19} = I \times t$

But number of moles of electrons = $2 \times \dfrac{m}{63.5}$

$$\Rightarrow L \times \text{Number of moles of electrons} \times 1.60 \times 10^{-19} = I \times t$$

$$\Rightarrow L \times 2 \times \frac{m}{63.5} \times 1.60 \times 10^{-19} = I \times t$$

$$\Rightarrow m = \frac{I \times t \times 63.5}{L \times 2 \times 1.60 \times 10^{-19}}.$$

Thus, if you keep the time duration (t) and current (I) constant and record mass readings for different quantities of current flow, you would be able to plot a graph of m versus quantity of charge, $Q = I \times t$. The gradient of the graph, $\frac{63.5}{L \times 2 \times 1.60 \times 10^{-19}}$, would enable you to determine Avogadro's constant. The way to tabulate the data is shown below:

Time/min	0	5	10	15	20	25	30
Current/A	1.0	1.0	1.0	1.0	1.0	1.0	1.0
Quantity of Charge/C	0	360	720	1080	1440	1800	2160
Mass of Cathode/g	20	20.12	20.24	20.36	20.48	20.60	20.72
Mass of Cu Deposited/g	0	0.12	0.24	0.36	0.48	0.60	0.72

Q But how is the procedure like?

A: Basically, after you have passed 1.0 A of current for 5 min, you stop the current. Then, you wash, dry, and weigh the cathode. After this, you place the cathode back and pass the same current for another 5 min and repeat the washing, drying, and weighing process. So, the mass is cumulative in nature as can be seen from the table above.

Q Now, instead of determining Avogadro's constant, can we use the same plan to determine Faraday's constant instead?

A: Yes, indeed. Faraday's constant is the quantity of charge ($Q = I \times t$) per mole of electrons:

$$L \times \text{Number of moles of electrons} \times 1.60 \times 10^{-19} = I \times t$$

$$\Rightarrow L \times 1.60 \times 10^{-19} = \frac{I \times t}{\text{Number of moles of electrons}} = \text{Faraday's constant}$$

Thus, experimentally, if you determine the quantity of charge by measuring the quantity of current that has passed through the circuit for a specific period of time and the number of moles of electrons indirectly from the electrolytic product, you would be able to determine Faraday's constant quite easily.

Note: There are many electrolytic experiments that you can perform in order to determine Avogadro's number and Faraday's constant:

(i) Electrolysis of molten lead(II) bromide, $PbBr_2$:

$$\text{At the anode: } 2Br^-(l) \rightarrow Br_2(l) + 2e^-$$
$$\text{At the cathode: } Pb^{2+}(l) + 2e^- \rightarrow Pb(s)$$

The lead metal formed at the cathode can be weighed to determine its mass and hence its number of moles. The rest of the calculations are similar to those above.

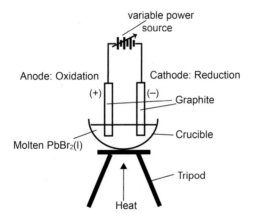

(ii) Electrolysis of water:

$$\text{At the anode: } 2H_2O(l) \rightarrow O_2(g) + 4H^+(aq) + 4e^-$$
$$\text{At the cathode: } 2H_2O(l) + 2e^- \rightarrow H_2(g) + 2OH^-(aq)$$

The O_2 and H_2 gases can be individually collected to determine their volumes and hence their number of moles. The rest of the calculations are similar to those above.

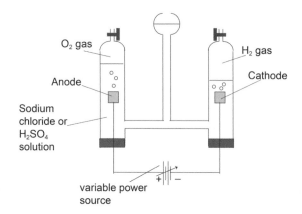

6.4 Design an Experiment for Copper Purification/ Design an Experiment to Electroplate an Object

The Task:

Copper metal that has been freshly extracted is loaded with numerous impurities. The presence of impurities increases the resistance of the metal. Hence, copper purification is a very important industrial process to improve the electrical conductivity of the copper metal.

Through the electrolysis of copper(II) sulfate solution using copper electrodes, plan an experiment to show how you can purify copper. Draw the experimental set-up. You are provided with the following chemicals and apparatus:

- 200 cm^3 of 1.00 mol dm^{-3} of CuSO$_4$ solution;
- A piece of pure copper metal;
- A piece of impure copper;
- Crocodile clips;
- Power supply;
- Ammeter; and
- Standard glassware in the lab.

> **Q** Where should we start thinking?

A: (i) What is the purpose of the plan?
— To purify copper.

(ii) What do you need to do in order to purify the impure copper?
— Place the impure copper as the anode, which when undergoing oxidation, will release copper as Cu^{2+} ions, leaving the impurities behind. Then, let the pure copper be the cathode which allows the Cu^{2+} ions to be reduced and deposited on it:

At the anode: $Cu(s) \rightarrow Cu^{2+}(aq) + 2e^-$

At the cathode: $Cu^{2+}(aq) + 2e^- \rightarrow Cu(s)$

Overall reaction: $Cu^{2+}(aq) + Cu(s) \rightarrow Cu^{2+}(aq) + Cu(s)$

> **Q** Is the set-up and procedure similar to those in Section 6.3?

A: Yes, they are. In fact, this same plan can be used for electroplating as well. In electroplating, the object to be plated is fixed as the cathode, which allows the metal ions to be reduced and deposited on it. A piece of pure metal is used as the anode to continuously supply metal ions into the electrolyte.

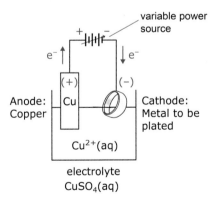

6.5 Design an Experiment to Anodize Aluminum

The Task:

Aluminum oxide is impervious to oxygen and water. This thus serves as a layer of protective oxide. Although the layer of aluminum oxide can be formed through oxidation, the process is slow to build up a substantial layer of the oxide. Therefore, the process of anodization is used to increase the protective surface coating on aluminum objects.

Through the electrolysis of sulfuric(VI) acid solution using graphite electrodes, plan an experiment to show how you can anodize an aluminum object with an oxide layer of 0.01 cm thickness and a density of 3.59 g cm^{-3}. Draw the experimental set-up. You are provided with the following chemicals and apparatus:

- 200 cm^3 of 1.00 mol dm^{-3} of H$_2$SO$_4$ solution;
- A piece of graphite electrode;
- A piece of aluminum object of surface area 25 cm^2;
- Crocodile clips;
- Power supply;
- Ammeter; and
- Standard glassware in the lab.

Q Where should we start thinking?

A: (i) What is the purpose of the plan?
— To anodize a piece of aluminum object.
(ii) What do you need to do in order to anodize the object?
— Make the object the anode, i.e., the reason why we call this process 'anodization,' where the aluminum metal will undergo oxidation.
(iii) But how can you determine the duration of anodization?
— Well, if we know the surface area of the object and the thickness of the oxide layer, based on the density of the aluminum oxide, we would be able to calculate the mass of the aluminum oxide. Then, calculate the number of moles of the Al$_2$O$_3$. With this number of moles, we would be able to determine the amount of electrons and hence the size of the current and the duration of the electrolytic process.

Q What is the purpose of the sulfuric acid?

A: The sulfuric acid is the electrolyte. We cannot simply use plain water as the electrolyte because the conductivity of plain water is very low. Thus, by using sulfuric acid as the electrolyte, the anodization process is essentially the electrolysis of water:

Anode (Al metal): Firstly, $2H_2O(l) \rightarrow O_2(g) + 4H^+(aq) + 4e^-$
Then, $4Al(s) + 3O_2(g) \rightarrow 2Al_2O_3(s)$
Cathode (Graphite): $2H^+(aq) + 2e^- \rightarrow H_2(g)$

Pre-Experimental Calculations:

Density of aluminum oxide = 3.59 g cm^{-3}

Given the surface area of the object is 25 cm^2 and the thickness of Al_2O_3 is 0.01 cm, volume of Al_2O_3 to be anodized = $25 \times 0.01 = 0.25$ cm^3

Mass of $Al_2O_3 = 0.25 \times 3.59 = 0.8975$ g

Molar mass of $Al_2O_3 = 2(27.0) + 3(16.0) = 102.0$ g mol^{-1}

Amount of $Al_2O_3 = \dfrac{0.8975}{102.0} = 8.80 \times 10^{-3}$ mol

Amount of $Al^{3+} = 2 \times$ Amount of $Al_2O_3 = 1.76 \times 10^{-2}$ mol

Amount of electrons released by
Al to $Al^{3+} = 3 \times$ Amount of $Al^{3+} = 5.28 \times 10^{-2}$ mol

Quantity of charge
= Amount of electrons released by Al to $Al^{3+} \times 96,500 = 5.095$ C

Assume that the current size is 0.1 A, the amount of time that is needed
$= \dfrac{5.095}{0.1} = 51$ s

Hence, let a current of 1 A flow for about 51 s.

> **Q** Is the set-up and procedure similar to those in Sections 6.3 and 6.4?

A: Yes, you are right. The differences are that (1) the cathode is a graphite rod while the anode is the aluminum object, and (2) the electrolyte is sulfuric acid.

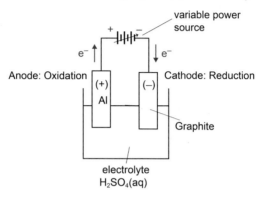

> **Q** If the Al_2O_3 is amphoteric, why didn't it react with the H^+ ions from the acid?

A: In electrolysis, the anode is positively charged as it is connected to the positive terminal of the power source. Hence, the H^+ ions would be repelled away.

6.6 Determine the Charge of an Ion Through Electrolysis

The Task:

Copper ions, Cu^{n+}, can be reduced through electrolysis as shown below:

$$Cu^{n+}(aq) + ne^- \rightarrow Cu(s).$$

Through the electrolysis of some Cu^{n+} ions, plan an experiment to show how you can determine the charge of the Cu^{n+} ions. Draw the experimental set-up. You are provided with the following chemicals and apparatus:

- 200 cm³ of 1.00 mol dm⁻³ of $Cu_2(SO_4)_n$ solution;
- A piece of pure copper metal;

- A piece of graphite rod;
- Crocodile clips;
- Power supply;
- Ammeter; and
- Standard glassware in the lab.

Q Where should we start thinking?

A: (i) What is the purpose of the plan?

— To determine the charge of the Cu^{n+} ions.

(ii) What do you need to do in order to determine the charge of the Cu^{n+} ions?

— We need to know how much of the copper has deposited at the cathode and how much of the current has passed through the circuit in a specific amount of time.

Pre-Experimental Calculations:

Let the mass of copper that has been deposited at the cathode be m g.
Let the time be 25 min = 1,500 s and the current be 1 A.

Amount of copper atoms = $\dfrac{m}{63.5}$ = a mol

Quantity of charge = $I \times t = 1 \times 1,500 = 1,500$ C

Amount of electrons = $\dfrac{I \times t}{96,500} = 1.55 \times 10^{-2}$ mol

$$Cu^{n+}(aq) + ne^- \rightarrow Cu(s)$$

Therefore, ratio of $Cu^{n+} : e^- = 1 : n =$ a $: 1.55 \times 10^{-2}$

$\Rightarrow \dfrac{1}{n} = \dfrac{a}{1.55 \times 10^{-2}}$

$\Rightarrow n = \dfrac{1.55 \times 10^{-2}}{a}$.

As the procedures are very similar to what we have discussed in Section 6.2, we would not discuss it further here.

Q What are the essentials in solving an electrolytic-planning task?

A: There are three main things that you need to think when solving an electrolytic-planning task: (1) the quantity (mass or volume or in number of moles) of substance that we can get from the anode, cathode, or both; (2) the amount of charge that has passed through the circuit which can be simply calculated using $Q = I \times t$; and (3) how to link up between the amount of substance that is "collected" at the electrode to the quantity of charge that has passed through the circuit.

6.7 Safety Precautions for Electrolytic Experiments

You may be asked to quote some safety precautions while performing an electrolytic experiment. Depending on the type of electrolytic experiment that you are performing, the following examples may be useful for you to take note:

— Handle electrical power of high voltage with care.
— Never allow an electrode to come in contact with any metal surfaces as it would cause an electrical short circuit.
— Before turning on the power source, make sure the voltage or current is set to zero. Then, turn on the voltage or current to your desired setting.
— Always wear gloves, a lab coat, and safety goggles while doing the experiment. For example, solid sodium hydroxide, acids (H_2SO_4) or bases (NaOH), methanol, etc. that you use may be corrosive in nature. So, there should be minimal direct contact of the skin with these chemicals.
— If there are any toxic gases or fumes evolved, conduct the experiment in a fumehood as such fumes may cause respiratory problems.
— If there are flammable liquids used, such as alcohol, ensure there is no naked flame around.
— Any organic chemicals should be properly covered or capped when not in use and used organic waste must be properly disposed.

6.8 Minimizing Experimental Errors or Increasing Reliability

You may be asked to quote some ways to improve experimental reliability while performing an electrolytic experiment. Depending on the type of electrolytic experiment that you are performing, the following examples may be useful for you to take note:

— The experiment must always be repeated to check for reliability of results. The average value should be calculated if necessary.
— You can refer to previous chapters regarding on how to improve reliability in the measurement of weight, volume, and other parameters.

CHAPTER 7

PLANNING FOR INORGANIC QUALITATIVE ANALYSIS

Qualitative analysis is a branch of Analytical Chemistry that involves knowing the *identity* of the compound only. If you have a mixture containing some anions or cations or both, how are you going to systematically determine each of their identities? How can you separate these ions from one another? Is there any specific test that would help you to pinpoint the identities of the species that are present?

Q What is considered a good chemical test in QA?

A: In QA, a good test must be one that allows a visual observation: color change, evolution of a gas, or formation of a precipitate.

Q What kinds of reactions would cause the formation of a precipitate?

A: Basically, a precipitate may form when two clear solutions (i.e., without the presence of a solid) are mixed. The formation of a precipitate indicates that the solid formed must be insoluble in water. So, you first need to know what are the insoluble compounds and the following table can help you to do that:

1. All sodium, potassium, ammonium, and nitrate compounds are soluble.
2. All carbonates and sulfates(IV) are insoluble in water, except for sodium carbonate, potassium carbonate, and ammonium carbonate.
3. All hydroxides are insoluble in water, except for Groups 1 and 2 (except Mg^{2+}). Calcium hydroxide is sparingly soluble.

4. Hydroxides of aluminum, chromium, lead, and zinc are soluble in excess NaOH(aq). Hydroxides of copper and zinc are soluble in excess NH_3(aq).
5. Barium sulfate, lead(II) sulfate, and calcium sulfate are insoluble in water.
6. Lead(II) halide and silver halide are insoluble in water.
7. All chromates except lead, barium, copper, iron(III), and silver are soluble.

Q Are there any other reactions that would lead to visual changes?

A: There are simply too many types of reactions; some might be acid–base in nature while others may be redox in nature:

Type of Reaction	Example
Precipitation	$2NaOH(aq) + Pb(NO_3)_2(aq) \rightarrow Pb(OH)_2(s) + 2NaNO_3(aq)$
	$Ag^+(aq) + Cl^-(aq) \rightarrow AgCl(s)$
Acid–base	$H_2SO_4(aq) + 2NaOH(aq) \rightarrow Na_2SO_4(aq) + 2H_2O(l)$
	$2H^+(aq) + CrO_4^{2-} \rightleftharpoons Cr_2O_7^{2-}(aq) + H_2O(l)$
Redox:	
Acid–metal	$2Na(s) + 2HCl(aq) \rightarrow 2NaCl(aq) + H_2(g)$
Combustion	$C(s) + O_2(g) \rightarrow CO_2(g)$
Displacement	$Cl_2(g) + 2KI(aq) \rightarrow I_2(aq) + 2KCl(aq)$
Disproportionation	$2H_2O_2(aq) \rightarrow 2H_2O(l) + O_2(g)$
Others	$2FeCl_2(aq) + Cl_2(g) \rightarrow 2FeCl_3(aq)$
Decomposition	$CuCO_3(s) \rightarrow CuO(s) + CO_2(g)$
	$2Cu(NO_3)_2(s) \rightarrow 2CuO(s) + 4NO_2(g) + O_2(g)$
	$Cu(OH)_2(s) \rightarrow CuO(s) + H_2O(g)$
Complex formation	Solid lead hydroxide dissolves in aqueous sodium hydroxide to form an aqueous complex salt called sodium plumbate:
	$Pb(OH)_2(s) + 2NaOH(aq) \rightarrow Na_2Pb(OH)_4(aq)$
	Silver chloride dissolves in aqueous NH_3 to form the diamminesilver(I) complex:
	$AgCl(s) + 2NH_3(aq) \rightarrow [Ag(NH_3)_2]^+(aq) + Cl^-(aq)$

Q Does the color of the unknown compound help us in the identification process?

A: Certainly! Certain compounds are colored because of the presence of a specific cation or anion. So, it is good to know some of their colors. The following table would be helpful:

	Possible Identity
Color of Solids/Precipitates	
Black	CuO, MnO_2, I_2, metal powder like Fe filings.
Blue	Copper(II) compounds, e.g., $CuSO_4$, $Cu(NO_3)_2$, and $Cu(OH)_2$.
Reddish-brown	$Fe(OH)_3$, copper metal.
Green	$CuCO_3$ and $CuCl_2$.
	$FeSO_4$ is pale green; $Fe(OH)_2$ is green.
	Chromium(III) compounds.
Orange	Potassium dichromate(VI), $K_2Cr_2O_7$.
Yellow	PbI_2 and AgI are pale yellow.
Purple	Potassium manganate(VII), $KMnO_4$.
White	Most compounds of Na^+, K^+, Ca^{2+}, Zn^{2+}, Al^{3+}, Pb^{2+}, NH_4^+.
	Important ones include $BaSO_4$, $PbSO_4$, $CaSO_4$, $PbCl_2$ and $AgCl$. $AgBr$ is off-white.
	{ZnO is yellow when hot and white when cooled.}
Color of Liquids/Solutions	
Blue	Aqueous copper(II) compounds.
Brown	Aqueous I_2 with aqueous KI.
Orange	Aqueous $K_2Cr_2O_7$.
	Bromine dissolves in organic solvent (e.g., hexane).
Yellow	Aqueous $FeCl_3$ (dark yellow).
	Aqueous Br_2.
	Aqueous I_2 (very pale yellow).
Purple	Aqueous $KMnO_4$.
Violet	Iodine dissolves in organic solvent (e.g., hexane) to form a violet solution.
Green	Aqueous chromium(III) compounds.
Pale green	Aqueous iron(II) compounds.

7.1 Sequence to Test for an Unknown Gas

If you suspect that a gas has evolved following a reaction, how can you sequentially identify the gas?

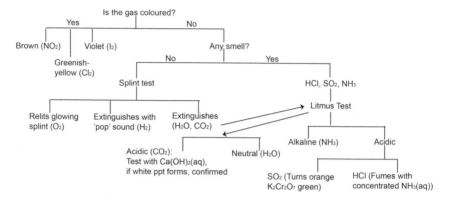

7.2 Identifying an Anion

First of all, you need to know the characteristic tests for each of the anion in the following table:

Anion	Characteristic Test	Observation
Carbonate (CO_3^{2-})	Add dilute acid.	Effervescence of CO_2 gas evolved. Gas turned moist blue litmus red and gave white ppt with $Ca(OH)_2$(aq).
Sulfate(IV) (SO_3^{2-})	Add dilute acid.	Gas turned moist blue litmus red and turned orange $K_2Cr_2O_7$ green.
	Add $Ba(NO_3)_2$(aq).	White ppt, $Ba(SO_3)_2$(s)
Chloride (Cl^-)	Acidify with dilute nitric acid, followed by aqueous silver nitrate, and then by NH_3(aq).	White ppt (AgCl) soluble in NH_3(aq).
	Or add lead(II) nitrate.	White ppt, $PbCl_2$(s).
Bromide (Br^-)	Acidify with dilute nitric acid, followed by aqueous silver nitrate, and then by NH_3(aq).	Off-white ppt (AgBr) partially soluble in NH_3(aq) but soluble in conc. NH_3(aq).
	Or add lead(II) nitrate.	White ppt, $PbBr_2$(s).

(Continued)

Planning for Inorganic Qualitative Analysis 225

(*Continued*)

Anion	Characteristic Test	Observation
Iodide (I$^-$)	Acidify with dilute nitric acid, followed by aqueous silver nitrate, and then by NH$_3$(aq).	Yellow ppt (AgI) insoluble in NH$_3$(aq).
	Or add lead(II) nitrate.	Yellow ppt, PbI$_2$(s).
Nitrate(V) (NO$_3^-$)	Add aqueous sodium hydroxide followed by Devarda's alloy (aluminum foil) and then warm gently.	Gas turned moist red litmus blue. Ammonia gas evolved.
Nitrate(III) (NO$_2^-$)	Add dilute acid.	Brown gas turned moist blue litmus red.
	Add aqueous sodium hydroxide followed by Devarda's alloy (aluminum foil) and then warm gently.	Gas turned moist red litmus blue. Ammonia gas evolved.
Sulfate (SO$_4^{2-}$)	Acidify with dilute nitric acid, followed by aqueous barium nitrate.	White ppt, BaSO$_4$(s).
Chromate(VI) (CrO$_4^{2-}$)	Add dilute acid.	Yellow solution turned orange with H$^+$(aq).
	Add Ba(NO$_3$)$_2$(aq) or Pb(NO$_3$)$_2$(aq)	Yellow ppt, BaCrO$_4$(s) or PbCrO$_4$(s).

Q Why must we acidify the sample with dilute nitric acid first before adding silver nitrate, lead(II) nitrate, or barium nitrate?

A: If there are any carbonate ions present in the sample, the addition of dilute nitric acid would decompose them. Hence, any precipitate that is subsequently formed would be due to the anion that is to be tested and not due to the carbonate ions.

Q Can we use H$_2$SO$_4$(aq) or HCl(aq) for the acidification?

A: No! This because H$_2$SO$_4$(aq) or HCl(aq) would introduce SO$_4^{2-}$ and Cl$^-$ ions, respectively. Thus, if there is a precipitate formed, it might be due to cations present in the sample which is able to form a precipitate with the SO$_4^{2-}$ or Cl$^-$ ions introduced. So, using nitric acid is a good choice because "all nitrates are soluble."

> **Q**: So, barium nitrate instead of barium chloride is used to test for SO_4^{2-} ions so as to prevent the introduction of Cl^- ions that may form a precipitate with cations in the sample?

A: Absolutely right!

To a solution that contains an unknown anion, you may carry out the following procedure in sequential order:

Step 1: Add dilute $H_2SO_4(aq)/HNO_3(aq)/HCl(aq)$ to a fresh sample.

```
                            Gas evolved   CO3²⁻, SO3²⁻, NO2⁻
Fresh sample → Add dilute ─┤
               H₂SO₄
           Yellow solution │ No gas
           turned orange   └─────────── I⁻, Cl⁻, Br⁻, NO3⁻, SO4²⁻
               CrO4²⁻
```

Step 2: Add $BaCl_2(aq)/Ba(NO_3)_2(aq)$ to a fresh sample.

```
                            White ppt ──── SO4²⁻
Fresh sample → Add dilute ─┤
               BaCl₂/
               Ba(NO₃)₂
                            No white ppt
                            ──────────── I⁻, Cl⁻, Br⁻, NO3⁻
```

Step 3: Add $AgNO_3(aq)$ to a fresh sample, followed by $NH_3(aq)$.

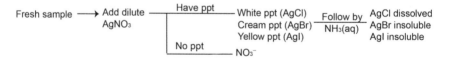

Step 4: Add $NaOH(aq)$ to a fresh sample and warm, followed by Devarda's alloy.

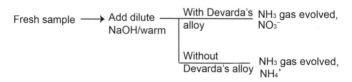

Q Why is it possible to use any of the three acids in Step 1? Does the anion of the acid matters?

A: Since our purpose is to test for the CO_3^{2-}, SO_3^{2-}, NO_2^-, or CrO_4^{2-} ion, the type of anion in the acid is not important. We are just making use of the H^+ ion from the acid. And anyway for each of the steps, we are using a fresh sample, so whatever reagent we have added in the previous step is not going to affect the next one.

7.3 Identifying a Cation

First of all, you need to know the characteristic tests for each of the cations in the following table:

Cation	Adding Aqueous Sodium Hydroxide	Adding Aqueous Ammonia
Aluminum (Al^{3+})	White ppt soluble in excess to give a colorless solution. $Al^{3+}(aq) + 3OH^-(aq) \rightarrow Al(OH)_3(s)$ $Al(OH)_3(s) + OH^-(aq) \rightarrow [Al(OH)_4]^-(aq)$	White ppt insoluble in excess. $Al^{3+}(aq) + 3OH^-(aq) \rightarrow Al(OH)_3(s)$
Ammonium (NH_4^+)	Ammonia evolved on warming. $NH_4^+(aq) + OH^-(aq) \rightarrow NH_3(g) + H_2O(l)$	—
Barium (Ba^{2+})	No ppt.	No ppt.
Calcium (Ca^{2+})	White ppt insoluble in excess. $Ca^{2+}(aq) + 2OH^-(aq) \rightarrow Ca(OH)_2(s)$ (Only with a concentrated solution.)	No ppt.
Chromium (Cr^{3+})	Grey-green ppt soluble in excess to give a dark green solution. $Cr^{3+}(aq) + 3OH^-(aq) \rightarrow Cr(OH)_3(s)$ $Cr(OH)_3(s) + 3OH^-(aq) \rightarrow [Cr(OH)_6]^{3-}(aq)$	Grey-green ppt insoluble in excess.

(Continued)

(*Continued*)

Cation	Adding Aqueous Sodium Hydroxide	Adding Aqueous Ammonia
Copper(II) (Cu^{2+})	Blue ppt insoluble in excess. $Cu^{2+}(aq) + 2OH^-(aq) \to Cu(OH)_2(s)$	Blue ppt soluble in excess to give a dark blue solution. $Cu^{2+}(aq) + 2OH^-(aq) \to Cu(OH)_2(s)$ $Cu(OH)_2(s) + 4NH_3(aq) \to [Cu(NH_3)_4]^{2+}(aq) + 2OH^-(aq)$
Iron(II) (Fe^{2+})	Green ppt insoluble in excess. $Fe^{2+}(aq) + 2OH^-(aq) \to Fe(OH)_2(s)$ Green ppt turned brown when exposed to air ($Fe(OH)_3$).	Green ppt insoluble in excess. $Fe^{2+}(aq) + 2OH^-(aq) \to Fe(OH)_2(s)$ Green ppt turned brown when exposed to air ($Fe(OH)_3$).
Iron(III) (Fe^{3+})	Reddish-brown ppt insoluble in excess. $Fe^{3+}(aq) + 3OH^-(aq) \to Fe(OH)_3(s)$	Reddish-brown ppt insoluble in excess. $Fe^{3+}(aq) + 3OH^-(aq) \to Fe(OH)_3(s)$
Lead(II) (Pb^{2+})	White ppt soluble in excess to give a colorless solution. $Pb^{2+}(aq) + 2OH^-(aq) \to Pb(OH)_2(s)$ $Pb(OH)_2(s) + 2OH^-(aq) \to [Pb(OH)_4]^{2-}(aq)$	White ppt insoluble in excess. $Pb^{2+}(aq) + 2OH^-(aq) \to Pb(OH)_2(s)$
Magnesium (Mg^{2+})	White ppt insoluble in excess. $Mg^{2+}(aq) + 2OH^-(aq) \to Mg(OH)_2(s)$	White ppt insoluble in excess. $Mg^{2+}(aq) + 2OH^-(aq) \to Mg(OH)_2(s)$
Manganese (Mn^{2+})	Off-white ppt insoluble in excess. $Mn^{2+}(aq) + 2OH^-(aq) \to Mn(OH)_2(s)$ The ppt turned brown upon standing in air. $Mn^{3+}(aq) + 3OH^-(aq) \to Mn(OH)_3(s)$	Off-white ppt insoluble in excess. $Mn^{2+}(aq) + 2OH^-(aq) \to Mn(OH)_2(s)$ The ppt turned brown upon standing in air. $Mn^{3+}(aq) + 3OH^-(aq) \to Mn(OH)_3(s)$
Zinc (Zn^{2+})	White ppt soluble in excess to give a colorless solution. $Zn^{2+}(aq) + 2OH^-(aq) \to Zn(OH)_2(s)$ $Zn(OH)_2(s) + 2OH^-(aq) \to [Zn(OH)_4]^{2-}(aq)$	White ppt soluble in excess to give a colorless solution. $Zn^{2+}(aq) + 2OH^-(aq) \to Zn(OH)_2(s)$ $Zn(OH)_2(s) + 4NH_3(aq) \to [Zn(NH_3)_4]^{2+}(aq) + 2OH^-(aq)$

Q Why does Ca^{2+} form a precipitate with aqueous NaOH but not with $NH_3(aq)$?

A: This is because $NH_3(aq)$ is a weaker base than aqueous NaOH. Hence, the concentration of the OH^- ion in $NH_3(aq)$ is too low to cause the $Ca(OH)_2$ to be precipitated out. For further details, you can refer to *Understanding Advanced Physical Inorganic Chemistry* by J. Tan and K. S. Chan.

To a solution that contains an unknown cation, you may carry out the following procedure in accordance to the flowchart below, starting with NaOH(aq):

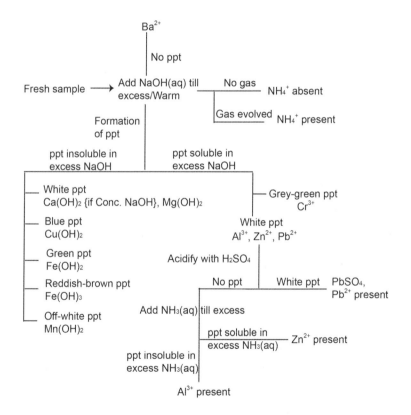

But if you prefer to start with NH₃(aq) as the testing reagent, then you may carry out the procedure in accordance to the following flowchart:

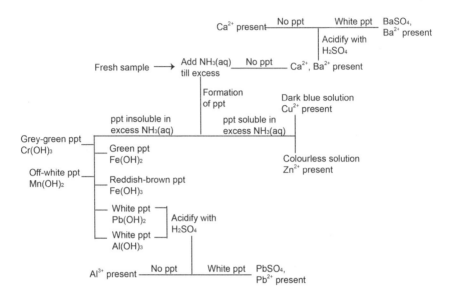

Take note that CaSO₄ can be precipitated out if the solution is highly concentrated.

7.4 Qualitative Analysis with an External Reagent

7.4.1 *Example* 1

> *The Task:*
>
> You are provided with four solutions labeled **FA1**, **FA2**, **FA3**, and **FA4**. They are aqueous solutions of aluminum chloride, barium hydroxide, zinc sulfate, and silver nitrate but are **not** arranged in the above order.
>
> You are to devise a **one-step** method to identify these solutions by using the solid sodium carbonate that is provided.
>
> You may not use any other chemicals or apparatus apart from the test tubes provided.

Solution:

(i) The unknown solution that produces effervescence of CO_2 gas with solid Na_2CO_3, must contain $AlCl_3(aq)$:

$$[Al(H_2O)_6]^{3+}(aq) + H_2O(l) \rightleftharpoons [Al(H_2O)_5(OH)]^{2+}(aq) + H_3O^+(aq);$$
$$2H_3O^+(aq) + Na_2CO_3(s) \rightarrow 2Na^+(aq) + CO_2(g) + 3H_2O(l).$$

(ii) Using $AlCl_3(aq)$, we can identify silver nitrate, as a white precipitate (AgCl) formed when these two unknown solutions react:

$$Ag^+(aq) + Cl^-(aq) \rightarrow AgCl(s).$$

(iii) As for the reaction between $Ba(OH)_2(aq)$ and $AlCl_3(aq)$, there would be a white precipitate $(Al(OH)_3)$ which is soluble in excess $Ba(OH)_2(aq)$, giving a colorless complex, $[Al(OH)_4]^-$:

$$Al^{3+}(aq) + 3OH^-(aq) \rightarrow Al(OH)_3(s);$$
$$Al(OH)_3(s) + OH^-(aq) \rightarrow [Al(OH)_4]^-(aq).$$

(iv) The other unknown solution that does not lead to any visible change with $AlCl_3(aq)$ is $ZnSO_4(aq)$. This can be confirmed by adding the $Ba(OH)_2(aq)$ that has been identified, to the $ZnSO_4(aq)$, you will get a white precipitate $(BaSO_4)$:

$$Ba^{2+}(aq) + SO_4^{2-}(aq) \rightarrow BaSO_4(s).$$

> **Q** Since $Ba(OH)_2(aq)$ can cause $Al(OH)_3(s)$ to dissolve in excess $Ba(OH)_2$, it is a strong base?

A: Yes, you are right.

> **Q** Does this mean that the Zn^{2+} of $ZnSO_4$ would also be soluble in excess $Ba(OH)_2(aq)$?

A: Absolutely spot on! The Zn^{2+} would form the $[Zn(OH)_4]^{2-}$ complex in excess $Ba(OH)_2(aq)$.

7.4.2 Example 2

The Task:

You are provided with four solutions labeled **FA1**, **FA2**, **FA3**, and **FA4**. They are aqueous solutions of sodium iodide, lead(II) nitrate, zinc sulfate, and potassium hydroxide but are **not** arranged in the above order.

You are to devise a **one-step** method to identify these solutions by using the aqueous hydrogen peroxide that is provided.

You may not use any other chemicals or apparatus apart from the test tubes provided.

Solution:

(i) The unknown solution that gives a brown solution ($I_2(aq)$) when $H_2O_2(aq)$ is added must contain NaI(aq):

$$H_2O_2(aq) + 2I^-(aq) + 2H^+(aq) \rightarrow I_2(aq) + 2H_2O(l).$$

(ii) Using NaI(aq), we can identify lead(II) nitrate, as a yellow precipitate (PbI_2) formed when these two solutions react:

$$Pb^{2+}(aq) + 2I^-(aq) \rightarrow PbI_2(s).$$

(iii) Add $Pb(NO_3)_2(aq)$ to the two remaining unknown solutions. The one that contains KOH(aq) will gives us a white precipitate ($Pb(OH)_2$) that is soluble in excess KOH(aq), forming a colorless complex, $[Pb(OH)_4]^{2-}$:

$$Pb^{2+}(aq) + 2OH^-(aq) \rightarrow Pb(OH)_2(s);$$

$$Pb(OH)_2(s) + 2OH^-(aq) \rightarrow [Pb(OH)_4]^{2-}(aq).$$

(iv) The other unknown solution that contains $ZnSO_4(aq)$ would give a white precipitate ($PbSO_4$) with the $Pb(NO_3)_2(aq)$:

$$Pb^{2+}(aq) + SO_4^{2-}(aq) \rightarrow PbSO_4(s).$$

7.4.3 Example 3

The Task:

You are provided with three colorless solutions labeled **FA1**, **FA2**, and **FA3**.

They are aqueous solutions of potassium iodide, sodium hydroxide, and lead(II) nitrate but are **not** necessary arranged in the above order.

By using only **one** of the reagents of sodium carbonate, potassium chromate(VI), or copper(II) sulfate, you can identify **FA1**, **FA2**, and **FA3**.

Explain which reagent you would use, and describe the observations which could help you identify the unknown solutions.

Solution:

(i) Copper(II) sulfate would be preferred because of the following observations that would be obtained:

— The test tube that gives a cream precipitate (CuI) and a brown solution (I_2(aq)) must contain potassium iodide:

$$2Cu^{2+}(aq) + 4I^-(aq) \rightarrow 2CuI(s) + I_2(aq).$$

— The test tube that gives a blue precipitate (Cu(OH)$_2$) must contain sodium hydroxide.

— The test tube that gives a white precipitate (a mixture of Pb(OH)$_2$ and PbCO$_3$) in blue solution must contain lead(II) nitrate.

(ii) Sodium carbonate is not preferred as it cannot differentiate between potassium iodide and sodium hydroxide.

(iii) Potassium chromate(VI) is not preferred as it cannot differentiate between potassium iodide and sodium hydroxide. In order for potassium chromate(VI) to oxidize the iodide ion, it must be an acidic medium.

Q Why is there a mixture of $Pb(OH)_2$ and $PbCO_3$ formed when sodium carbonate is added to lead (II) nitrate?

A: The carbonate ion, CO_3^{2-} undergoes basic hydrolysis to produce hydroxide ions, OH^-. Thus, both the carbonate and hydroxide salts would precipitate out. Take note, if the cation is triply charged such as Fe^{3+}, Al^{3+} or Cr^{3+}, only the hydroxide is precipitated out. This is because the carbonate ions would be decomposed by the acid that is produced from the appreciable hydrolysis of the triply charges cation. For details, refer to *Understanding Advanced Physical Inorganic Chemistry* by J. Tan and K. S. Chan.

7.5 Self-Contained Qualitative Analysis

Self-contained QA refers to QA tasks which make use of one of the unknown solutions as a testing reagent, without the "help" of other external testing reagents.

7.5.1 *Example 1*

The Task:

You are provided with five solutions labeled **FA1**, **FA2**, **FA3**, **FA4**, and **FA5**. They are aqueous solutions of lead(II) nitrate, zinc nitrate, copper(II) sulfate, sodium hydroxide, and sodium dichromate(VI) but are **not** arranged in the above order.

You are to devise a **one-step** method to identify these solutions by choosing one of them, dividing it into four portions, and adding a portion to each of the four remaining solutions.

The chosen solution may be regarded as the **reagent**.

You may not use any other chemicals or apparatus apart from the test tubes provided.

State the **reagent used** and outline how it might be used to differentiate the other solutions.

Solution:

(i) Out of the five unknowns, copper(II) sulfate is blue in color while sodium dichromate(VI) is orange in color. So, these two solutions can be easily identified.

(ii) Then, add the sodium dichromate(VI) to each of the remaining three unknown solutions:
- The test tube that gives an orange precipitate (PbCr$_2$O$_7$) must contain lead(II) nitrate.
- The test tube that gives an reddish-brown precipitate (CuCr$_2$O$_7$) must contain copper(II) sulfate.
- The test tube that causes the orange sodium dichromate(VI) solution to turn yellow must contain sodium hydroxide:

$$Cr_2O_7^{2-}(aq) + 2OH^-(aq) \rightleftharpoons H_2O(aq) + 2CrO_4^{2-}.$$

- The test tube that does not lead to any visible change must contain zinc nitrate, which is originally colorless.

Q Can we use the copper(II) sulfate solution as the testing reagent instead?

A: Yes, of course. The following would be the expected results:
- The test tube that gives a white precipitate (PbSO$_4$) in a blue solution must contain lead(II) nitrate.
- The test tube that gives an reddish-brown precipitate (CuCr$_2$O$_7$) must contain sodium dichromate(VI).
- The test tube that gives a blue precipitate (Cu(OH)$_2$), must contain sodium hydroxide.
- The test tube that does not lead to any visible change must contain zinc nitrate, which is originally colorless.

7.5.2 Example 2

The Task:

You are provided with four colorless solutions labeled **FA1**, **FA2**, **FA3**, and **FA4**.

They are aqueous solutions of hydrochloric acid, sodium carbonate, lead(II) nitrate, and aluminum chloride but are **not** necessary arranged in the above order.

Without any indicators and using only these solutions alone, plan the steps which will enable you to identify these solutions.

Solution:

As none of the solution is colored, the only way is to mix the solutions with one another and then tabulate the visible changes (precipitate formation or effervescence) in a table as shown below:

	FA1	FA2	FA3	FA4
FA1				
FA2				
FA3				
FA4				

— The unknown solution which gives effervescence of CO_2 gas with two of the other unknown solutions, and produce a white precipitate with the third unknown solution, must contain $Na_2CO_3(aq)$:

$$2HCl(aq) + Na_2CO_3(aq) \rightarrow 2NaCl(aq) + CO_2(g) + H_2O(l);$$

$$[Al(H_2O)_6]^{3+}(aq) + H_2O(l) \rightleftharpoons [Al(H_2O)_5(OH)]^{2+}(aq) + H_3O^+(aq);$$

$$2H_3O^+(aq) + Na_2CO_3(aq) \rightarrow 2Na^+(aq) + CO_2(g) + 3H_2O(l);$$

$$Pb(NO_3)_2(aq) + Na_2CO_3(aq) \rightarrow PbCO_3(s) + 2NaNO_3(aq).$$

— Of the two unknown solutions that give effervescence of CO_2 with $Na_2CO_3(aq)$, one of them does not form a white precipitate with $Na_2CO_3(aq)$. This unknown solution is $HCl(aq)$. The one that gives a white precipitate with $Na_2CO_3(aq)$ is $AlCl_3(aq)$:

$$Al^{3+}(aq) + 3OH^-(aq) \rightarrow Al(OH)_3(s).$$

— The unknown solution which gives a white precipitate with all the other three unknown solutions must contain $Pb(NO_3)_2(aq)$:

$$Pb^{2+}(aq) + 2HCl(aq) \rightarrow PbCl_2(s) + 2H^+(aq);$$

$$3Pb^{2+}(aq) + 2AlCl_3(aq) \rightarrow 3PbCl_2(s) + 2Al^{3+}(aq);$$

$$Pb^{2+}(aq) + Na_2CO_3(aq) \rightarrow PbCO_3(s) + 2Na^+(aq).$$

Treatment of Results:

	FA1	FA2	FA3	FA4
FA1		White ppt	Effervescence	Effervescence
FA2	White ppt		White ppt	White ppt
FA3	Effervescence	White ppt		No visible reaction
FA4	Effervescence	White ppt	No visible reaction	

FA1 is $Na_2CO_3(aq)$; **FA2** is $Pb(NO_3)_2(aq)$; **FA3** is $HCl(aq)$; and **FA4** is $AlCl_3(aq)$.

> **Q** When $AlCl_3(aq)$ reacts with $Na_2CO_3(aq)$, the white precipitate is $Al(OH)_3$ and not $Al_2(CO_3)_2$. Why is it so? Where does the OH^- ion come from?

A: The CO_3^{2-} in water undergoes basic hydrolysis:

$$CO_3^{2-}(aq) + H_2O(l) \rightleftharpoons HCO_3^-(aq) + OH^-(aq).$$

Thus, $CO_3^{2-}(aq)$ or $HCO_3^-(aq)$ or $OH^-(aq)$ can actually react with the $H_3O^+(aq)$ produced from the hydrolysis of the Al^{3+} ion. But you can imagine that "all the $CO_3^{2-}(aq)$ has been converted to $HCO_3^-(aq)$ and $OH^-(aq)$" and the $H_3O^+(aq)$ reacts "only" with the $HCO_3^-(aq)$. At the end of the day, there is only $OH^-(aq)$ left and the OH^- can form a precipitate with the Al^{3+} ion. Or, from another perspective, there are three stages of hydrolysis for the Al^{3+} ion:

$$[Al(H_2O)_6]^{3+}(aq) + H_2O(l) \rightleftharpoons [Al(H_2O)_5(OH)]^{2+}(aq) + H_3O^+(aq);$$

$$[Al(H_2O)_5(OH)]^{2+}(aq) + H_2O(l) \rightleftharpoons [Al(H_2O)_4(OH)_2]^+(aq) + H_3O^+(aq);$$

$$[Al(H_2O)_4(OH)_2]^+(aq) + H_2O(l) \rightleftharpoons [Al(H_2O)_3(OH)_3](s) + H_3O^+(aq).$$

Thus, when $Na_2CO_3(aq)$ is added, the CO_3^{2-} ion reacts with the $H_3O^+(aq)$ formed; this drives the positions of equilibrium for the three equations till the stage of forming the insoluble $Al(OH)_3$.

7.5.3 Example 3

The Task:

You are provided with six solutions labeled **FA1**, **FA2**, **FA3**, **FA4**, **FA5**, and **FA6**.

They are aqueous solutions of potassium chromate(VI), sulfuric acid, barium nitrate, sodium chloride, potassium iodide, and lead(II) nitrate but are **not** necessary arranged in the above order.

Without any indicators and using only these solutions alone, plan the steps which will enable you to identify these solutions.

Solution:

(i) The test tube that contains a yellow solution is aqueous potassium chromate(VI).

(ii) Then, add the aqueous potassium chromate(VI) to the rest of the five unknown solutions:

— The test tube that causes the yellow potassium chromate(VI) to turn orange must contain sulfuric acid:

$$2H^+(aq) + CrO_4^{2-} \rightleftharpoons Cr_2O_7^{2-}(aq) + H_2O(l).$$

— There will be unknown two solutions that give a yellow precipitate with potassium chromate(VI); they are barium nitrate and lead(II) nitrate:

$$Pb^{2+}(aq) + CrO_4^{2-}(aq) \rightarrow PbCrO_4(s);$$

$$Ba^{2+}(aq) + CrO_4^{2-}(aq) \rightarrow BaCrO_4(s).$$

— The other two unknown solutions that do not give a precipitate with potassium chromate(VI); they are potassium iodide and sodium chloride.

— Mix the two unknown solutions that give a yellow precipitate (barium nitrate and lead(II) nitrate) with the other two unknown solutions that do not give a precipitate (potassium iodide and sodium chloride) with potassium chromate(VI):

— The unknown solution that gives a yellow precipitate (PbI_2) and white precipitate ($PbCl_2$) is lead(II) nitration solution;

— The unknown solution that gives a white precipitate (PbCl$_2$) with one unknown solution only is sodium chloride solution;
— The unknown solution that gives a yellow precipitate (PbI$_2$) with one unknown solution only is potassium iodide solution; and
— The unknown solution that does not give any precipitate is barium nitrate solution.

7.5.4 Example 4

The Task:

You are provided with six solutions labeled **FA1**, **FA2**, **FA3**, **FA4**, **FA5**, and **FA6**.

They are aqueous solutions of potassium bromide, magnesium nitrate, barium nitrate, zinc sulfate, potassium iodide, and lead(II) nitrate but are **not** necessary arranged in the above order.

Without any indicators and using only these solutions alone, plan the steps which will enable you to identify these solutions.

Solution:

As none of the solution is colored, the only way is to mix the solutions with one another and then tabulate the visible changes (precipitate formation or effervescence) in a table as shown below:

	FA1	FA2	FA3	FA4	FA5	FA6
FA1						
FA2						
FA3						
FA4						
FA5						
FA6						

— The unknown solution that gives two white precipitates and a yellow precipitate with three other unknown solutions, must contain $Pb(NO_3)_2$(aq):

$$Pb(NO_3)_2(aq) + 2KBr(aq) \rightarrow PbBr_2(s) + 2KNO_3(aq);$$

$$Pb(NO_3)_2(aq) + ZnSO_4(aq) \rightarrow PbSO_4(s) + Zn(NO_3)_2(aq);$$

$$Pb(NO_3)_2(aq) + 2KI(aq) \rightarrow PbI_2(s) + 2KNO_3(aq).$$

From here, both the identities of $Pb(NO_3)_2$(aq) and KI(aq) can be confirmed.

— The unknown solution that gives two white precipitate with two other unknown solutions must contain zinc sulfate:

$$Ba(NO_3)_2(aq) + ZnSO_4(aq) \rightarrow BaSO_4(s) + Zn(NO_3)_2(aq);$$

$$Pb(NO_3)_2(aq) + ZnSO_4(aq) \rightarrow PbSO_4(s) + Zn(NO_3)_2(aq).$$

Now, once the $ZnSO_4$(aq) is confirmed, this would mean that the KBr(aq) is also confirmed.

— Hence, out of the two remaining solutions, $Ba(NO_3)_2$(aq) or $Mg(NO_3)_2$(aq), the unknown solution that gives a white precipitate with $ZnSO_4$(aq) must contain $Ba(NO_3)_2$(aq):

$$Ba(NO_3)_2(aq) + ZnSO_4(aq) \rightarrow BaSO_4(s) + Zn(NO_3)_2(aq).$$

Treatment of Results:

	FA1	FA2	FA3	FA4	FA5	FA6
FA1		White ppt	No visible reaction	No visible reaction	No visible reaction	No visible reaction
FA2	White ppt		No visible reaction	No visible reaction	White ppt	No visible reaction
FA3	No visible reaction	No visible reaction		No visible reaction	No visible reaction	No visible reaction
FA4	No visible reaction	No visible reaction	No visible reaction		White ppt	No visible reaction
FA5	No visible reaction	White ppt	No visible reaction	White ppt		Yellow ppt
FA6	No visible reaction	No visible reaction	No visible reaction	No visible reaction	Yellow ppt	

FA1 is $Ba(NO_3)_2$(aq); **FA2** is $ZnSO_4$(aq); **FA3** is $Mg(NO_3)_2$(aq); **FA4** is KBr(aq); **FA5** $Pb(NO_3)_2$(aq); and **FA6** KI(aq).

7.5.5 Example 5

> *The Task:*
>
> You are provided with five solutions labeled **FA1**, **FA2**, **FA3**, **FA4**, and **FA5**. They are aqueous solutions of aluminum sulfate, barium nitrate, lead(II) nitrate, potassium iodide, and iron(III) nitrate but are **not** necessary arranged in the above order. Without any indicators and using only these solutions alone, plan the steps which will enable you to identify these solutions.

Solution:

(i) The test tube that contains a yellow solution is iron(III) nitrate.
(ii) Then, add the aqueous iron(III) nitrate solution to the rest of the four unknown solutions:
— The unknown solution that gives a brown solution (I_2(aq)), must contain potassium iodide:

$$2Fe^{3+}(aq) + 2I^-(aq) \rightarrow 2Fe^{2+}(aq) + I_2(aq).$$

(iii) Then, add the aqueous potassium iodide solution to the rest of the three unknown solutions:
— The unknown solution that gives a yellow precipitate (PbI_2), must contain lead(II) nitrate.
(iv) Then, add the aqueous lead(II) nitrate solution to the rest of the two unknown solutions:
— The unknown solution that gives a white precipitate ($PbSO_4$), must contain aluminum sulfate.
(v) The last unknown solution left must be barium nitrate. You can confirm it by adding the aluminum sulfate that has been identified in step (iv), and you will see a white precipitate of $BaSO_4$ formed.

7.6 Separation of Ions
7.6.1 *Example 1*

The Task:

You are given an aqueous solution containing three cations: Pb^{2+}, Al^{3+}, and Zn^{2+} ions. You are to devise a method to separate the three cations so that they can be obtained separately as **precipitates**. The reagents provided are:

- Aqueous sodium hydroxide;
- Aqueous sodium carbonate;
- Aqueous ammonia;
- Dilute nitric acid; and
- Dilute hydrochloric acid.

Plan a sequence of steps and write down the expected observations and locations where the cations can be obtained in a table format.

Solution:

Step	Expected Observation	Location of Cation
1. Add dilute HCl in excess to the mixture to precipitate out Pb^{2+} as $PbCl_2$.	White ppt formed.	$PbCl_2(s)$ $Zn^{2+}(aq)$, $Al^{3+}(aq)$
Filter the mixture and wash the residue.	White residue. Colorless filtrate.	$PbCl_2(s)$ $Zn^{2+}(aq)$, $Al^{3+}(aq)$
2. To the filtrate from step (1) add $NH_3(aq)$ in excess.	White ppt formed, some soluble in excess $NH_3(aq)$.	$Al(OH)_3(s)$, $Zn(OH)_2(s)$ $[Zn(NH_3)_4]^{2+}(aq)$
Filter the mixture and wash the residue.	White residue. Colorless filtrate.	$Al(OH)_3(s)$ $[Zn(NH_3)_4]^{2+}(aq)$
3. Add dil HNO_3 dropwise to the filtrate from step (2).	White ppt formed.	$Zn(OH)_2(s)$
Filter the mixture and wash the residue.	White residue. Colorless filtrate.	$Zn(OH)_2(s)$

7.6.2 Example 2

The Task:

You are given an aqueous solution containing: Cu^{2+}, Fe^{3+}, Cl^-, and SO_4^{2-} ions. You are to devise a method to separate these ions so that they can be obtained separately as **precipitates**. The reagents provided are:

- Aqueous sodium hydroxide;
- Aqueous silver nitrate;
- Aqueous ammonia;
- Dilute nitric acid; and
- Aqueous barium nitrate.

Plan a sequence of steps and write down the expected observations and locations where the ions can be obtained in a table format.

Solution:

Step	Expected Observation	Location of Cation
1. Add aqueous sodium hydroxide to the mixture.	Blue ppt formed insoluble in excess NaOH(aq).	$Cu(OH)_2(s)$
	Reddish-brown ppt formed, insoluble in excess NaOH(aq).	$Fe(OH)_3(s)$ $Cl^-(aq)$, $SO_4^{2-}(aq)$
Filter the mixture and wash the residue.	Reddish-brown and blue residue.	$Cu(OH)_2(s)$, $Fe(OH)_3(s)$
Retain the filtrate for step (5).	Colorless filtrate.	$Cl^-(aq)$, $SO_4^{2-}(aq)$
2. To the residue from step 1, add dilute HNO_3 to dissolve the residue.	Residue dissolved to give a green solution.	$Cu^{2+}(aq)$–Blue $Fe^{3+}(aq)$–Yellow (Blue + Yellow → Green)
3. Add $NH_3(aq)$ till excess to the solution from step (2).	Reddish-brown ppt formed insoluble in excess. Blue ppt formed soluble in excess to give a dark blue solution.	$Fe(OH)_3(s)$ $[Cu(NH_3)_4]^{2+}(aq)$
Filter the mixture and wash the residue. Retain the filtrate for step (4).	Reddish-brown residue. Dark blue filtrate.	$Fe(OH)_3(s)$ $[Cu(NH_3)_4]^{2+}(aq)$

(Continued)

(Continued)

Step	Expected Observation	Location of Cation
4. To the filtrate from step (3), add dilute HNO$_3$.	Blue ppt formed.	Cu(OH)$_2$(s)
Filter the mixture and wash the residue.	Blue residue. Colorless filtrate.	Cu(OH)$_2$(s)
5. To the filtrate from step (1), add dilute HNO$_3$ to neutralize the excess NaOH, followed by the addition of Ba(NO$_3$)$_2$(aq).	White ppt formed.	BaSO$_4$(s) Cl$^-$(aq)
Filter the mixture and wash the residue. Retain the filtrate for step (5).	White residue. Colorless filtrate.	BaSO$_4$(s) Cl$^-$(aq)
6. To the filtrate from step (5), add AgNO$_3$(aq).	White ppt formed.	AgCl (s)
Filter the mixture and wash the residue.	White residue. Colorless filtrate.	AgCl(s)

7.6.3 Example 3

The Task:

You are given an aqueous solution containing: Cl$^-$, Br$^-$, and I$^-$ ions. You are to devise a method to separate the three anions so that they can be obtained separately as **precipitates**. The reagents provided are:

- Concentrated ammonia;
- Aqueous silver nitrate;
- Aqueous ammonia; and
- Dilute nitric acid.

Plan a sequence of steps and write down the expected observations and locations where the anions can be obtained in a table format.

Solution:

Step	Expected Observation	Location of Cation
1. Add aqueous silver nitrate to the mixture.	White ppt formed. Cream ppt formed. Yellow ppt formed.	$AgCl(s)$ $AgBr(s)$ $AgI(s)$
Filter the mixture and wash the residue.	Yellowish-white residue. Colorless filtrate.	$AgCl(s), AgBr(s), AgI(s)$
2. To the residue from step (1), add dilute $NH_3(aq)$ to dissolve the residue.	White residue dissolved. Cream and yellow residues were insoluble.	$[Ag(NH_3)_2]^+, Cl^-(aq)$ $AgBr(s), AgI(s)$
Filter the mixture and wash the residue. Retain the filtrate for step (4).	Yellow-white residue. Colorless filtrate.	$AgBr(s), AgI(s)$ $[Ag(NH_3)_2]^+, Cl^-(aq)$
3. To the residue from step (3), add conc. NH_3.	Cream residue dissolved. Yellow residue was insoluble.	$[Ag(NH_3)_2]^+, Br^-(aq)$ $AgI(s)$
Filter the mixture and wash the residue. Retain the filtrate for step (5).	Yellow residue. Colorless filtrate.	$AgI(s)$ $[Ag(NH_3)_2]^+, Br^-(aq)$
4. To the filtrate from step (2), add dilute HNO_3.	White ppt formed.	$AgCl(s)$
Filter the mixture and wash the residue.	White residue. Colorless filtrate.	$AgCl(s)$
5. To the filtrate from step (3), add dilute HNO_3.	Cream ppt formed.	$AgBr(s)$
Filter the mixture and wash the residue.	Cream residue. Colorless filtrate.	$AgBr(s)$

7.7 Safety Precautions for Qualitative Analysis

You may be asked to quote some safety precautions while performing qualitative analysis. Depending on the type of qualitative analysis that you are performing, the following examples may be useful for you to take note:

— Always wear gloves, a lab coat, and safety goggles while doing the experiment. For example, solid sodium hydroxide, acids (H_2SO_4) or bases (NaOH), methanol, etc. that you use may be corrosive in nature.

So, there should be minimal direct contact of the skin with these chemicals.
- If there are any toxic gases or fumes evolved, conduct the experiment in a fumehood as such fumes may cause respiratory problems.
- When heating a test tube of solution, move the test tube in the flame. Do not localize the heating as it would cause the solution to boil suddenly and shoot out from the test tube.
- Ensure that there are no organic chemicals lying around when there is a naked flame.
- Use a windshield to surround the flame. This can prevent the hot solution from shooting out from the test tube during heating and injuring others.

CHAPTER 8

PLANNING FOR ORGANIC QUALITATIVE ANALYSIS

Qualitative analysis can also be used for organic compounds possessing a specific functional group that result in characteristic reactions and leading to visible changes. If you are given a few unknown organic compounds, each possessing a specific functional group, how are you going to systematically determine each of their identities? Are there any specific tests that would help you to pinpoint the identities of the functional groups that are present?

8.1 Common Organic Functional Group Tests

First of all, you need to know the characteristic tests for each of the functional group in the following table. Next, always use simple chemical tests that would lead to visible changes. If hydrolysis needs to be done, for esters and amides, then strong heating is required. Otherwise, all warming must be done in a water bath for safety reasons. For detailed chemical reactions and balanced equations, please refer to *Understanding Advanced Organic and Analytical Chemistry* by K. S. Chan and J. Tan.

Functional Group	Characteristic Test	Observation
Alkene $\diagdown_{C=C}\diagup$	(i) Br_2/CCl_4 or $Br_2(aq)$. (ii) Hot acidified $KMnO_4$. (iii) Cold alkaline $KMnO_4$.	(i) Reddish-brown Br_2 decolorized. (ii) Purple $KMnO_4$ decolorized. (iii) Purple $KMnO_4$ decolorized, brown MnO_2 ppt formed.

(*Continued*)

(*Continued*)

Functional Group	Characteristic Test	Observation
Alkylbenzene except R—C(R)(R)—C₆H₅ (tertiary carbon attached to benzene ring)	(i) Hot acidified $KMnO_4$. (ii) Hot alkaline $KMnO_4$.	(i) Purple $KMnO_4$ decolorized. (ii) Purple $KMnO_4$ decolorized, brown MnO_2 ppt formed.
Halogenoalkane, RX where X = Cl, Br or I	Add NaOH(aq) to each of these compounds and heat. Next, cool the mixture followed by adding HNO_3(aq). Lastly, add $AgNO_3$(aq).	For RCl, white ppt formed (AgCl). For RBr, cream ppt formed (AgBr). For RI, yellow ppt formed (AgI).
Alcohol, ROH	(i) Add Na metal. (ii) Hot acidified $KMnO_4$ (except for tertiary alcohols). (iii) Hot alkaline $KMnO_4$ (except for tertiary alcohols). (iv) Hot acidified $K_2Cr_2O_7$ (except for tertiary alcohols). (v) I_2/NaOH/warm (only for alcohols of the type $CH_3CH(OH)R$, where R is H, an alkyl group, or an aryl group). (vi) Use PCl_5.	(i) Effervescence of H_2 gas. (ii) Purple $KMnO_4$ decolorized. (iii) Purple $KMnO_4$ decolorized, brown MnO_2 ppt formed. (iv) Orange $K_2Cr_2O_7$ turned green. (v) Yellow ppt (CHI_3) formed. (vi) White fumes of HCl.
Phenol (C₆H₅—OH)	(i) Add Na metal. (ii) Add aqueous NaOH. (iii) Aqueous $FeCl_3$. (iv) Aqueous Br_2.	(i) Effervescence of H_2 gas. (ii) Phenol dissolved. Two layers became one. (iii) Violet complex formed. (iv) Reddish-brown Br_2 decolorized. White ppt formed ($C_6H_2Br_3OH$).

(*Continued*)

(Continued)

Functional Group	Characteristic Test	Observation
Aldehyde, RCHO	(i) Hot acidified $KMnO_4$. (ii) Hot alkaline $KMnO_4$. (iii) Hot acidified $K_2Cr_2O_7$. (iv) I_2/NaOH/warm (only for aldehyde of the type CH_3CHO). (v) 2,4-DNPH. (vi) Tollens's reagent ($[Ag(NH_3)_2]^+$)/warm. (vii) Fehling's reagent (Cu^{2+} complex)/warm. (Note that Fehling's reagent does not respond to aromatic aldehydes.)	(i) Purple $KMnO_4$ decolorized. (ii) Purple $KMnO_4$ decolorized, brown MnO_2 ppt formed. (iii) Orange $K_2Cr_2O_7$ turned green. (iv) Yellow ppt (CHI_3) formed. (v) Orange ppt formed. (vi) Silver mirror formed. (vii) Reddish-brown ppt formed.
Ketone, RCOR	(i) I_2/NaOH/warm (only for ketones of the type CH_3COR). (ii) 2,4-DNPH.	(i) Yellow ppt (CHI_3) formed. (ii) Orange ppt formed.
Carboxylic acid, RCOOH	(i) Add Na metal. (ii) Add NaOH(aq). (iii) Add Na_2CO_3(aq). (iv) Use PCl_5.	(i) Effervescence of H_2 gas. (ii) Acid dissolved. Two layers became one. (iii) Effervescence of CO_2 gas. (iv) White fumes of HCl.
Acid halide, RCOX	(i) Add water. (ii) Add water to each of these compounds. Next, add $AgNO_3$(aq).	(i) Acid fumes (HCl, HBr, HI) observed. (ii) For RCOCl, white ppt formed (AgCl). For RCOBr, cream ppt formed (AgBr).

(Continued)

(Continued)

Functional Group	Characteristic Test	Observation
Ester, RCOOR	(i) Hot acidified $KMnO_4$. (ii) Hot alkaline $KMnO_4$. (iii) Hot acidified $K_2Cr_2O_7$.	(i) Purple $KMnO_4$ decolorized. (ii) Purple $KMnO_4$ decolorized, brown MnO_2 ppt formed. (iii) Orange $K_2Cr_2O_7$ turned green. (The acid hydrolyzed the ester to give an alcohol, which was oxidized by $KMnO_4$ or $K_2Cr_2O_7$. Tertiary alcohols cannot be oxidized.)
Amide, $RCONH_2$	Add NaOH(aq), heat strongly. Test gas evolved with moist red litmus.	$NH_3(g)$ (or other amine vapors) evolved turned moist red litmus blue.
Amine	Moist red litmus.	Moist red litmus turned blue (amine is basic).
Phenylamine ⌬—NH_2	(i) Moist red litmus. (ii) Aqueous Br_2.	(i) Moist red litmus turned blue. (ii) Reddish-brown Br_2 decolorized. White ppt formed ($C_6H_2Br_3NH_2$).
Ammonium (NH_4^+) salt	Add NaOH(aq), warm gently. Test gas evolved with moist red litmus.	$NH_3(g)$ evolved turned moist red litmus blue.
Nitrile, $RC \equiv N$	Add NaOH(aq), heat strongly. Test gas evolved with moist red litmus.	$NH_3(g)$ evolved turned moist red litmus blue.

8.2 Identifying an Organic Functional Group

8.2.1 *Example 1*

The Task:

You are provided with six aqueous solutions of organic compounds, labeled **FA1**, **FA2**, **FA3**, **FA4**, **FA5**, and **FA6**. They are:

- Ethanol, CH_3CH_2OH;
- Propanal, CH_3CH_2CHO;
- Propanone, CH_3COCH_3;
- Ethanoic acid, CH_3COOH;
- Phenol, C_6H_5OH; and
- Ethyl ethanoate, $CH_3COOCH_2CH_3$.

Outline a sequence of simple chemical tests, by which you could identify each of the above organic substances. You are not allowed to identify the substances by elimination.

Solution:

(i) Add aqueous Na_2CO_3 to each of the five test tubes. The one that produces effervescence of CO_2 gas that gives a white precipitate with $Ca(OH)_2$, must contain ethanoic acid, CH_3COOH.

(ii) Add Tollens's reagent to the remaining five test tubes and warm. The one that gives a silver mirror ($Ag(s)$) must contain propanal, CH_3CH_2CHO.

OR

Add Fehling's reagent to the remaining five test tubes and warm. The one that gives a reddish-brown precipitate ($Cu_2O(s)$) must contain propanal, CH_3CH_2CHO.

(iii) Add 2,4-DNPH to each of the four remaining test tubes. The one that gives an orange precipitate must contain propanone, CH_3COCH_3.

(iv) Add aqueous $FeCl_3$ solution to each of the three remaining test tubes. The one that gives a violet coloration must contain phenol, C_6H_5OH.

OR

Add aqueous Br_2 solution to each of the three remaining test tubes. The one that gives a white precipitate and decolorizes reddish-brown bromine must contain phenol, C_6H_5OH.

(v) Add a piece of sodium metal to each of the two remaining test tubes. The one that gives effervescence of H_2 gas which extinguishes a lighted splint with a 'pop' sound, must contain ethanol, CH_3CH_2OH.

OR

Add aqueous iodine with sodium hydroxide (I_2/NaOH) to each of the two remaining test tubes with warming. The one that gives a yellow precipitate (CHI_3) must contain ethanol, CH_3CH_2OH.

OR

Add PCl_5 to each of the two remaining test tubes. The one that gives white fumes of HCl gas must contain ethanol, CH_3CH_2OH.

(vi) Add hot acidified $KMnO_4$ to the last test tube. If it decolorizes the purple $KMnO_4$, it must contain ethyl ethanoate, $CH_3COOCH_2CH_3$. The ethanol that is formed from the hydrolysis of the ester is oxidized by the acidified $KMnO_4$.

OR

Add hot alkaline $KMnO_4$ to the last test tube. If it decolorizes the purple $KMnO_4$ and gives a brown precipitate (MnO_2) it must contain ethyl ethanoate, $CH_3COOCH_2CH_3$. The ethanol formed from the hydrolysis of the ester is oxidized by the alkaline $KMnO_4$.

OR

Add hot acidified $K_2Cr_2O_7$ to the last test tube. If the mixture changes from orange to green (Cr^{3+}(aq)), it must contain ethyl ethanoate, $CH_3COOCH_2CH_3$. The ethanol formed from the hydrolysis of the ester is oxidized by the orange acidified $K_2Cr_2O_7$.

Q What is the acid that used to acidify $KMnO_4$ or $K_2Cr_2O_7$?

A: We use H_2SO_4 to acidify as HCl can be oxidized by $KMnO_4$. You can use HCl to acidify $K_2Cr_2O_7$ as it is a weaker oxidizing agent than $KMnO_4$.

8.2.2 Example 2

> **The Task:**
>
> You are provided with five aqueous solutions of organic compounds, labeled **FA1**, **FA2**, **FA3**, **FA4**, and **FA5**. They are:
>
> - $CH_3CH(OH)COOH$;
> - $CH_3COCOOH$;
> - $HOOCCOCH_2CH_2COOH$;
> - $CH_3CH=CHCOOH$; and
> - CH_3CH_2COOH.
>
> Outline a sequence of simple chemical tests, by which you could identify each of the above organic substances.

Solution:

(i) Add aqueous iodine in NaOH to each of the five test tubes, and warm the mixture. The test tubes that give a yellow precipitate (CHI_3) contain one of the following two compounds:

$$CH_3CH(OH)COOH \text{ or } CH_3COCOOH.$$

Then, add 2,4-DNPH to each of the two above solutions. The one that gives an orange precipitate contains $CH_3COCOOH$. The other one that does not give an orange precipitate must contain $CH_3CH(OH)COOH$.

OR

Then, add hot acidified $KMnO_4$ to each of the two above solutions. The one that decolorizes the purple $KMnO_4$ contains $CH_3CH(OH)COOH$. The remaining one that does not decolorize purple $KMnO_4$ contains $CH_3COCOOH$.

(ii) To the three remaining test tubes, add 2,4-DNPH. The one that gives an orange precipitate contains $HOOCCOCH_2CH_2COOH$. The other two that do not give an orange precipitate must contain CH₃CH=CHCOOH and CH₃CH₂COOH.

(iii) Now, add aqueous Br_2 (or Br_2 in CCl_4) to the last two test tubes. The one that decolorizes reddish-brown Br_2 contains $CH_3CH=CHCOOH$. The remaining one that does not decolorize reddish-brown Br_2 contains CH_3CH_2COOH.

OR

Add hot acidified $KMnO_4$ to the last two test tubes. The one that decolorizes purple $KMnO_4$ contains $CH_3CH=CHCOOH$. The remaining one that does not decolorize purple $KMnO_4$ contains CH_3CH_2COOH.

8.2.3 Example 3

The Task:

You are provided with four organic compounds, labeled **FA1**, **FA2**, **FA3**, and **FA4**. They are:

- Ethanoyl chloride, CH_3COCl;
- Ethanoyl bromide, CH_3COBr;
- (Chloromomethyl)benzene, $C_6H_5CH_2Cl$; and
- Chlorobenzene, C_6H_5Cl.

Outline a sequence of simple chemical tests, by which you could identify each of the above organic substances.

Solution:

(i) Add aqueous silver nitrate to each of the four test tubes. The test tube that immediately gives a white precipitate (AgCl) must contain ethanoyl chloride, CH_3COCl. The one that immediately gives a cream precipitate (AgBr) must contain ethanoyl bromide, CH_3COBr. The other two test tubes would not have any visible observations.

(ii) To the remaining two test tubes:

Add NaOH(aq) to each of these compounds and heat. Next, cool the mixture followed by adding HNO_3(aq). Lastly, add $AgNO_3$(aq). The test tube that gives a white precipitate (AgCl) must contain (chloromomethyl)benzene, $C_6H_5CH_2Cl$. The one that does not give any precipitate must contain chlorobenzene, C_6H_5Cl.

Q Why was NaOH(aq) not added before adding the aqueous silver nitrate in step (i)?

A: The acid halides are very reactive to nucleophilic attacks. As such, the water molecules in aqueous silver nitrate can act as a nucleophile to substitute the halide ion easily.

Q Then, can we simply add water without the silver nitrate?

A: Although the adding of water would result in the formation of acid fumes, both of the acid fumes, HCl and HBr, are colorless. Thus, you cannot differentiate between the acid chloride from the acid bromide.

Q Can we use lead(II) nitrate in place of silver nitrate?

A: Why not? The only issue that you may have is being unable to differentiate between $PbCl_2$ and $PbBr_2$, which are both white.

Q Why does the chlorobenzene not give a white precipitate of AgCl?

A: This is because the lone pair of electrons on the Cl atom delocalizes into the benzene ring. This results in a partial double bond character between the Cl atom and the carbon atom. Hence, it is difficult for chlorobenzene to undergo nucleophilic substitution.

Now, take note that if the Cl group is bonded next to an alkene double bond (C = C), there is also partial double bond character between the Cl atom and the alkene carbon atom (refer to *Understanding Advanced Organic and Analytical Chemistry* by K. S. Chan and J. Tan).

8.3 Safety Precautions for Organic Qualitative Analysis

You may be asked to quote some safety precautions while performing organic qualitative analysis. Depending on the type of organic qualitative analysis that you are performing, the following examples may be useful for you to take note:

— Always wear gloves, a lab coat, and safety goggles while doing the experiment. For example, solid sodium hydroxide, acids (H_2SO_4) or bases (NaOH), methanol, etc. that you use may be corrosive in nature. So, there should be minimal direct contact of the skin with these chemicals.
— If there are any toxic gases or fumes evolved, conduct the experiment in a fumehood as such fumes may cause respiratory problems.
— If there are flammable liquids used, such as alcohol, ensure there is no naked flame around.
— Always use a hot water-bath for heating and warming.
— Any organic chemicals should be properly covered or capped when not in use and used organic waste must be properly disposed.

CHAPTER 9

PLANNING FOR ORGANIC SYNTHESIS

Complex organic compounds can be synthesized from simple molecules via organic reactions. The reaction process can be a single step or multiple steps. It is quite unlikely for an organic synthetic reaction to give us 100% yield of the product without the formation of side products. As such, separation and purification techniques are essential in organic synthesis. A typical organic synthetic procedure would involve: (1) the mixing of reactants in a reactor; (2) supply of heat energy if the reaction is slow or cooling the reactor down if the reaction is highly exothermic; (3) separation of products as usually not all the reactants are converted to the desired product; (4) purification of the product as even the desired product that has been separated would still contain impurities; and lastly (5) confirmation of the identity of the product through simple methods such as melting/boiling point determination to the more complicated usage of high-end instrumentations such as X-ray crystallography, infrared spectroscopy, and nuclear magnetic resonance (NMR) spectroscopy.

9.1 Starting the Organic Reaction

The reactants are usually mixed in a conical flask or a round-bottomed flask:

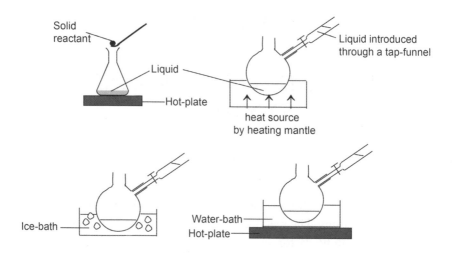

If heat energy needs to be supplied, it is usually done through a hot plate or a heating mantle that fits the capacity of the round-bottomed flask. A 100-ml round-bottomed flask cannot be heated by a heating mantle which is meant for a 150-ml round-bottomed flask as the 100-ml round-bottomed flask cannot be heated evenly.

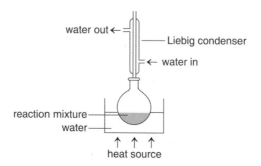

If prolonged heating is required, a Liebig condenser needs to be attached to the round-bottomed flask as shown. This is called heating under reflux. The term 'reflux' implies continuous boiling with condensation taking place. This allows for prolonged heating without the contents drying up too fast. But if the product formed needs to be immediately distilled out when formed, we would need to use the heat-with-immediate distillation set-up. This is very similar to a normal simple distillation set-up (see Section 9.2.1).

If cooling is necessary because the reaction is highly exothermic, then the round-bottomed flask can be cooled in an ice bath or under running water and the reactant should be added slowly with continuous stirring.

9.2 Separation of Products

9.2.1 *Separation of miscible liquids*

If the contents after the reaction consist of liquids of different boiling points, then separation can be effected by using simple distillation or fractional distillation. If the two liquids have a boiling point difference of more than about 25°C, the two liquids can be separated via *simple distillation*. Or, if we need to recover a solvent that has been used to dissolve a solid, for example, to recover the water from seawater, we can also use this simple set-up:

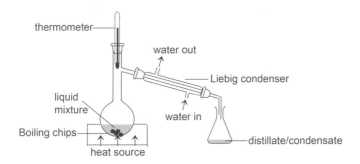

Q: If we have two liquids of different boiling points, how would we know that we have fully separated the two liquids?

A: Firstly, we need to know that when a liquid boils, the boiling point is constant. So, when you heat up the mixture, the liquid with a lower boiling point would boil at a particular temperature and the temperature would stay constant. Until when all the liquid with a lower boiling point has evaporated, you would see an increase in the temperature corresponding to the boiling point of the other liquid. You can stop heating at this juncture as the two liquids have been separated.

Q What is the purpose of the boiling chips?

A: The boiling chips are to ensure smooth boiling by letting small vapor bubbles form within the porous chips. If boiling chips are not added, vapor bubbles cannot form and the liquid may be heated beyond its boiling point. This is called 'superheating.' At this higher temperature, large bubbles are formed suddenly and would explode. There would be "bumping" in the liquid.

Q Why is the thermometer placed at the entrance where the vapor exits into the Liebig condenser?

A: Before the liquid starts boiling, the temperature is much lower than that in the liquid mixture as there is no vapor to heat up the thermometer. But when the liquid starts boiling and its vapor rises, the temperature of the vapor is going to be similar to that of the boiling liquid. Thus, when you see a constant temperature while the liquid is boiling, you know that you are collecting what you want.

Q Why is the cooling water entering the Liebig condenser at the bottom rather than at the top?

A: By letting the cooling water enter the condenser at the bottom, it ensures that at this end, the temperature is the coolest. This would condense all the vapor without any loss. But if the two liquids have a difference in boiling point of less than about 25°C, then we have to use fractional distillation to separate the two liquids. Fractional distillation can be used to separate a mixture of more than two liquids. The *closer* the boiling points of the different liquids, the *longer* the fractionating column we need to use, so it is not an issue.

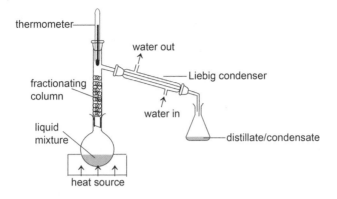

> **Q** How does the fractionating column help to separate two or more liquids?

A: When you heat the liquids, the various components vaporize. The components that have higher boiling points would condense at the lower end of the column and fall back to the flask again. The ones that have lower boiling points would continue to rise and condense at an upper point in the column. Amongst these components, those that have slightly higher boiling points would condense while the rest continue to move up. So, basically, at each point of the column, the temperature corresponds to the boiling point of a particular liquid, which means that multiple simple distillations take place at different points of the column.

> **Q** Why is the column packed with glass beads?

A: The glass beads provide a greater surface area for the vapor to condense.

> **Q** How would the temperature readings on the thermometer be like?

A: First, the temperature reading would correspond to that of the lowest boiling point component. When all these have vaporized and condensed, then the thermometer would indicate the boiling point of the next component with a higher boiling point, so on and so forth, until the last component is vaporized.

> **Q** How would you remove the small amount of water in a solvent? You can't carry out a distillation process, right?

A: If there is still some water in the solvent after distillation, what you can do is to add some anhydrous reagents such as anhydrous $MgSO_4$ or Na_2SO_4 into the "wet" solvent. These ionic compounds will "extract" the water molecules from the solvent through the formation of strong ion-dipole interactions (refer to *Understanding Advanced Physical Inorganic Chemistry* by J. Tan and K. S. Chan). You can then perform a filtration to remove the drying agent.

9.2.2 Separation of two immiscible liquids

If we have two liquids, such as water and oil, that do not mix with each other and are separated into two distinct layers, we say that they are *immiscible*. A separating/separatory/tap funnel is used to separate such a liquid mixture.

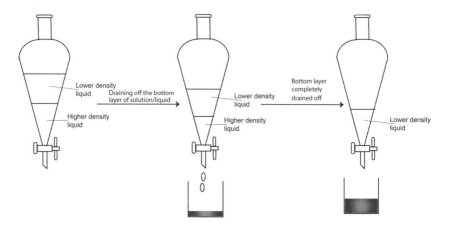

This technique is particularly useful in separating organic products from an aqueous reaction mixture. The reaction mixture is shaken with successive amounts of a water-immiscible organic solvent. When the mixture is shaken in the separating funnel, the pressure in the funnel should be periodically released. This is to avoid pressure build-up in the separating funnel, especially if a gas is produced in the process. The pressure build-up will cause the stopper to be pushed out and product might be lost. Each time, the organic layer is run off and collected before fresh solvent is again added. The organic material will distribute itself between the aqueous medium and the organic solvent until equilibrium is reached, but more will dissolve in the organic layer. Two or three extractions with a suitable solvent will remove most of the product.

Q: Let's say we need to extract a particular desired product from a plethora of other side products. Is there a way to check which two phases are optimal for the isolation of a particular component of interest?

A: Yes. Basically, the differential distribution of a particular substance **X** over two different phases is characterized by an unique equilibrium constant known as partition coefficient or distribution coefficient, $P_{1/2}$, which is defined as follows:

$$P_{1/2} = \frac{\text{concentration of substance X in phase 1}}{\text{concentration of substance X in phase 2}}.$$

A large $P_{1/2}$ indicates that the interaction of substance **X** with the particles of phase 1 is more favorable than those of substance **X** with the particles of phase 2. Like all other equilibrium constants, the partition coefficient is temperature dependent. Thus, what one needs to do is to compare our desired product with something that has already been documented and try using the documented desired phases. But, if one really cannot find something that is similar, then one probably have to resort using the trial-and-error method if the physical property is known, such as the polarity, of the substance of interest.

Take for instance, in the preparation of phenylamine from the reduction of nitrobenzene using tin with concentrated hydrochloric acid:

$$C_6H_5NO_2 + Sn(s)/\text{conc. HCl} \rightarrow C_6H_5NH_3Cl.$$

The phenylamine would be in the protonated form and hence soluble in water. What we can do is to add some sodium hydroxide, which will convert the protonated phenylamine into the unprotonated form:

$$C_6H_5NH_3^+ + OH^- \rightarrow C_6H_5NH_2 + H_2O.$$

We can then use another less polar solvent to extract the phenylamine using the separating funnel. Similarly, in the oxidation of methylbenzene ($C_6H_5CH_3$) to benzoic acid (C_6H_5COOH) using alkaline $KMnO_4$, the product will be the benzoate ion ($C_6H_5COO^-$). Acidification can be done first before extraction using the separatory funnel:

$$C_6H_5COO^- + H^+ \rightarrow C_6H_5COOH + H_2O.$$

> **Q**: If both of the two immiscible layers are colorless, how would know which layer is which?

A: Well, you can use the density data to help you. If not, if one of the layer is aqueous, all you need to do is take a dropper and drop in few drops of water. If the water droplets do not "travel through" the top layer, then this top layer is the aqueous layer. But if you see the droplets moving through the top layer to the bottom layer, then the bottom layer must be the aqueous layer.

9.2.3 Separation of a solid from a liquid

A simple filtration using a filter funnel can help to separate a solid from a liquid, but the filtration process is slow as we need the help of gravity to "pull" the liquid through the filter funnel. To accelerate the filtrating process, Buchner filtration can be performed. Essentially, the filter flask is connected to a vacuum pump which creates a partial vacuum environment within the flask. The high atmospheric pressure above the surface of the liquid mixture pushes the liquid through the filter paper.

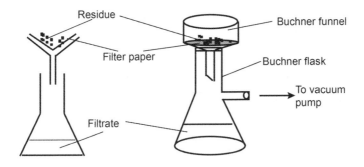

In addition, the sucking of the air through the residue can also help us to air dry the residue faster.

Q What should we do if it is the filtrate that we want to isolate?

A: Well, retain the filtrate and carry out extraction process on the filtrate. If not, just evaporate the filtrate to dryness using a water bath.

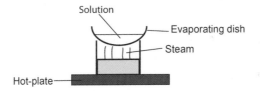

9.2.4 Separation using chromatography

You would probably have come across chromatography as a technique used in separating component dyes in black ink. The basic procedure first involves using a pencil to draw a baseline on a rectangular strip of chromatography paper. Sample black ink is spotted onto the baseline and the entire paper is then inserted upright into a beaker containing a suitable solvent, making sure the baseline is above the solvent level. Immediately, the solvent creeps up the paper by *capillary action* and following its trail, are colored spots moving at different speeds. Once the solvent is near the edge of the paper, the paper is removed and the solvent front is marked out. The developed paper with the spots in their final positions is called the *chromatogram*.

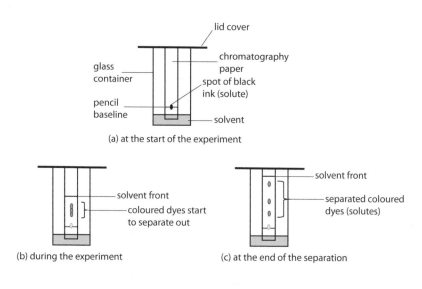

(a) at the start of the experiment

(b) during the experiment

(c) at the end of the separation

Q: What is capillary action?

A: Capillary action refers to the phenomenon whereby liquid level rises through a *thin* narrow tube of a small diameter or through porous materials such as cloth or paper. If one looks closely at a piece of cloth or paper, one would notice that it is made up of multiple small little capillaries or "channels." The rising of the liquid level is against gravity and it is possible, because of strong interactive forces between the liquid molecules and the surfaces of the substance.

Q: Why does it happen only for capillary tubes? What happens if the diameter of the tube is too big?

A: If the diameter of the tube is too big, then as the molecules cling onto the wall at the side, the weight of the molecules at the center would pull it down. In a capillary tube, because it is too narrow, there aren't *many* molecules at the center of the tube, hence the weight to pull down the molecules at the side is not too great. Therefore, the molecules can "climb" up the wall.

Q Why is a lid placed over the container?

A: The container is covered with a lid so as to ensure that a constant and stable vapor atmosphere encloses the chromatogram. If not, the evaporation of the solvent would create a draft and this would interfere with the movement of the solvent front and the separation of the different types of solute. In addition, the solvent molecules at the solvent front would also evaporate at different rates. This would cause a non-uniform solvent front to be obtained, which poses problems in the calculation of the retardation factor or retention factor.

Q What is the retardation factor or retention factor?

A: The retardation factor is the ratio of the distance traveled by the solute to the distance traveled by the solvent as shown below:

$$R_f = \frac{\text{distant traveled by solute}}{\text{distant traveled by solvent}}.$$

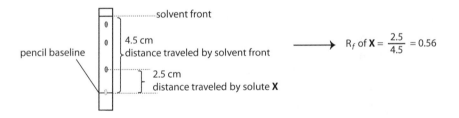

This is why it is important to mark the solvent front immediately at the end of the experiment, before it evaporates from the paper. Each substance has a different R_f value. The unknown substance can be identified by matching its R_f value obtained against a database of R_f values of some known compounds. As the R_f values are obtained experimentally, there are a few factors that we need to look into that would affect the data obtained. These factors include the type of paper used, the solvent and its concentration, the amount of solutes being spotted, and the temperature.

> **Q** Can we use a pen to draw the baseline?

A: If we use a pen to draw the baseline, the dyes from the ink would separate as the solvent front moves.

> **Q** Why must the pencil baseline be above the solvent level?

A: If the baseline is below the solvent level, the sample would dissolve in the solvent.

The principle behind the technique lies in the different extents of adsorption of the solutes in each of the two phases needed to carry out the experiment. The solutes are a mixture of molecules that you intend to separate and the solvent in the beaker is the mobile phase in which the solutes can dissolve. That is why the solutes (the spots) are observed to "travel" across the paper. But some solutes travel the least whereas the rest the farthest away from the baseline. Why do these solutes move across different distances? This is where we bring in the idea of the second phase.

The second phase is the stationary phase across which the mobile phase moves. The two phases differ in their nature of polarity. So, what we have here is a solute that has a higher affinity for one phase than the other. If a solute dissolves better in the mobile phase, it will be "carried" farther across the chromatography paper. If it has a higher affinity for the stationary phase, it will reside longer at a particular spot, opposing the "tidal force" of the mobile phase as the solvent moves forward. Hence, choosing an appropriate solvent system is crucial for the success of a separation process using chromatography.

Take for instance, for the separation of two solutes (as shown below), hexane is used as the moving phase, as shown by the chromatogram on the left. But if ethanol is used, the chromatogram for the separation is the one in the middle. This shows that the two solutes are relatively more soluble in the polar ethanol than the non-polar hexane. Hence, the two solutes "prefer" to move along with the solvent front.

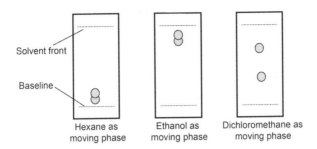

So, to effectively separate the two solutes, a slightly less-polar solvent, such as dichloromethane may be used, as shown by the chromatogram on the right.

Q What is the stationary phase in paper chromatography?

A: In paper chromatography, the stationary phase consists of the water molecules that are embedded within the porous structure of the cellulose fibers. Hence, this stationary phase is polar in nature as the cellulose is actually made up of poly(glucose). This would mean that solute particles that are polar in nature would have a greater affinity for this polar water/cellulose stationary phase, and would be adsorbed more strongly onto the stationary phase. Therefore, it would be "retained" farther behind the solvent front. In contrast, if the solute is less polar and thus interacts much better with the non-polar mobile phase instead, then we would expect this less-polar solute to move more in line with the solvent front. Thus, such differential interactions with the mobile and stationary phases allow different solutes to be separated.

Celluslose – A Poly(glucose)

Q So, if the stationary phase is polar, does it mean that we have to use an absolutely non-polar mobile phase?

A: Not really. If the stationary phase is polar, we can still employ a polar mobile phase. Importantly, there must be a substantial difference in the polarities of the two different phases to allow a significant differential distribution of the solute to take place between these two phases. For example, we can use the polar ethanol as the mobile phase since ethanol is relatively less polar than the water/cellulose stationary phase.

 Q How should we write the procedure for a chromatographic experiment for the separation of compounds?

A: Well, a general procedure for a chromatographic separation process is as follows:

(1) Fill a jar with an appropriate solvent about 1-cm deep. This is the moving phase. Cover the jar to allow a saturated vapor environment to build up.
(2) Use a pencil to gently draw the baseline on a piece of filter paper or alumina-silica plate as shown in p. 266.
(3) Spot the sample compound mixture and other known compounds on the baseline, well separated from each other.
(4) Use a forceps to introduce the paper gently into the jar of solvent. Make sure that the solvent level is below the pencil baseline.
(5) When the solvent front is near the top of the paper, remove the paper from the jar.
(6) Quickly use a pencil to mark the solvent front on the chromatogram.

 Q Why must we use a forceps to introduce the paper into the jar? Can we use our hands to handle it?

A: It is not wise to use your hands as they might contain contaminants which can be transferred onto the filter paper.

9.3 Purification of Compounds

9.3.1 *Method of sublimation*

If we have a mixture where one of the components can be sublimed (i.e., it transforms from the solid to gaseous state without passing through the liquid state), e.g., caffeine, naphthalene, iodine, ammonium chloride, and solid carbon dioxide, then we can make use of the following procedure:

1. Heat the mixture in an evaporating dish using a hot water bath or an infrared lamp.
2. Invert a filter funnel as shown below to allow the gaseous compound to solidify on the cooler surface of the funnel:

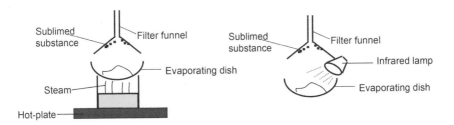

9.3.2 *Method of recrystallization*

Recrystallization is only used if the desired product is a solid and not a liquid. In the recrystallization methodology, both the desired compound and impurities are dissolved in a suitable solvent at a higher temperature. It is assumed that the quantities of the impurities are minute and the desired compound dissolves better at the higher temperature. With a higher solubility at the elevated temperature, more of the compound can be dissolved in a small volume of solvent. This would mean that when the solution is slowly cooled down, the desired compound would crystallize out, leaving the impurities behind in the solvent. The mixture is then filtered to retain the residue which is the desired compound.

But if the impurities are insoluble in the solvent, then the mixture can be dissolved using the hot solvent through the filter paper. The dissolved compound is collected, leaving the impurities on the filter funnel. Later on, the filtrate is cooled down to maximize the formation of the solid crystals.

Now, if you are asked to write down a typical set of steps for recrystallization, the procedure is as follows:

1. Boil some solvent in a separate conical flask. Place the mixture onto the filter paper.

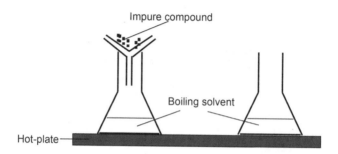

2. Ensure that the whole filtration set-up is heated up by the solvent vapor before pouring in the hot solvent to dissolve the compound. Use as little amount of solvent as possible.

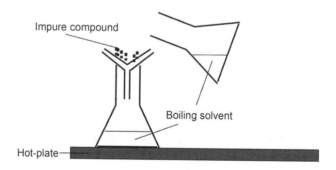

3. After all the compound has dissolved, saturate the solution by boiling off some of the solvent.

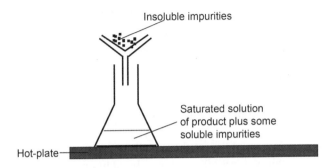

4. Cool the saturated solution slowly to room temperature or below room temperature using an ice bath.

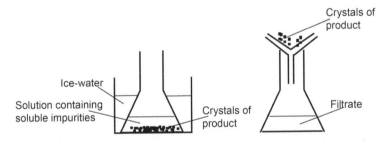

5. Filter the crystals and air-dry it or dry it by using an infrared lamp. Repeat the recrystallization process if necessary to produce as pure a product as possible.

Q Why must we let the solvent heat up the filtration set-up first?

A: If the filtration set-up is not heated up, the moment the hot solution passes through the cooler filter funnel, the dissolved compound is going to crystallize out on the cooler surface. This would affect the yield of the product.

> **Q** How would you know that you have a saturated solution?

A: When you start to see crystals form during the boiling off of the solvent, this would mean that you do not have a sufficient amount of solvent to dissolve the compound. So, what you can do is used a dropper to add some hot solvent just to dissolve the crystals.

> **Q** Why must we cool the saturated solution slowly?

A: Too fast a cooling would cause the crystals to form suddenly. This may trap impurities within the crystal structures. In addition, the size of the crystal that is formed is small during rapid cooling. Such small crystals may not be useful for further analysis such as X-ray crystallography.

> **Q** How do you air-dry the compound? Just leave it in the air?

A: No. You can use Buchner filtration to air-dry it or use an infrared lamp to warm up the wet compound.

> **Q** So, is it alright to recrystallize as many times as possible in order to get a much purer product?

A: No! There must be a balance between purity and yield. Each recrystallization process would result in the loss of the desired compound. Too many recrystallizations would result in a smaller amount of yield but a purer compound.

9.4 Testing for the Purity of a Substance

It is very important to know if the compound that we have isolated is in the pure state. For example, if a drug contains impurities that are harmful to the body, then it is not consumable.

But after you have isolated a compound, how do you know whether it is in the pure form? The presence of impurities would *decrease* the melting point of a solid. This is because the presence of "extraterrestrial" particles in the original solid crystal lattice structure *disrupts* the *regular arrangement* of the particles. This thus affects the *strength* of the bond, lowering the melting point.

The presence of impurities in a liquid increases the boiling point of a liquid. This is because the particles of the impurities *attract* the original randomly moving liquid particles closer together. Thus, more energy is needed to overcome this *stronger* attractive force.

Hence, we can test whether a compound is pure or not by determining the melting or boiling point and compare it against that of the pure form in a database.

Melting point determination:

A small sample of the dry solid is placed in a melting-point tube and the temperature is raised very gradually. The temperature at which the solid melts, i.e., the temperature when the crystalline solid structure collapses and a meniscus is formed in the sample tube, is recorded as the melting point of the solid.

9.5 Safety Precautions for Organic Synthesis

You may be asked to quote some safety precautions while performing organic synthesis. Depending on the type of organic synthesis that you are performing, the following examples may be useful for you to take note:

— Always wear gloves, a lab coat, and safety goggles while doing the experiment. For example, solid sodium hydroxide, acids (H_2SO_4) or bases (NaOH), methanol, etc. that you use may be corrosive in nature. So, there should be minimal direct contact of the skin with these chemicals.
— If there are any toxic gases or fumes evolved, conduct the experiment in a fumehood as such fumes may cause respiratory problems.
— If there are flammable liquids used, such as alcohol, ensure there is no naked flame around. All heating must be done using a water bath. But if a higher temperature is needed, then a heating mantle is preferred.
— Any organic chemicals should be properly covered or capped when not in use and used organic waste must be properly disposed.

CHAPTER 10

PLANNING FOR SPECTROPHOTOMETRIC ANALYSIS

White light is a polychromatic electromagnetic radiation consisting of several monochromatic electromagnetic radiations of different colors. Each monochromatic radiation has a specific frequency and wavelength related by the following equation:

$c = \lambda.f,$ where c = speed of the electromagnetic wave in m s^{-1},

λ = wavelength in m, and

f = frequency in s^{-1}.

The longer the wavelength of the color, the less energetic is the radiation.

When white light comes in contact with an object, some of the light may be reflected or absorbed or transmitted through the object. If the object reflects all the white light, the object will look white in color. Vice versa, if the object absorbs all the white light, the object looks black. But if the object absorbs some of the wavelengths of the transmitted visible light, then the color that is observed will be complementary to the color that is being absorbed. Meaning? A red object looks red because the complementary green color has been absorbed.

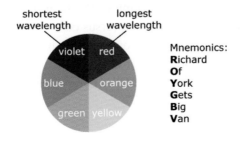

Thus, by knowing the wavelength of the light absorbed by a particular compound, one can make use of the amount of radiation absorbed (absorbance, A) to determine the concentration of the compound through the Beer–Lambert Law:

$$A = \log_{10}(I_0/I) = \varepsilon.c.l,$$ where
- A = absorbance,
- I_0 = intensity of radiation going into the sample,
- I = intensity of radiation coming out from the sample,
- ε = the molar extinction coefficient ($dm^3\ mol^{-1}\ cm^{-1}$),
- c = concentration of compound (in $mol\ dm^{-3}$), and
- l = path length of the absorbing solution (in cm).

The concentration of a particular compound in a solution is proportional to the intensity of the radiation absorbed (A). The molar extinction coefficient reflects how strongly the compound is able to absorb the radiation *per se* and it is an intrinsic property of the compound. The higher the molar extinction coefficient of a compound, the smaller the amount of the compound needed for analysis. To quantitatively determine the concentration of a compound in question, simply follow the following procedure:

1. Use a spectrophotometer to determine the visible spectrum of the compound in question, and of other compounds that may be present in the mixture by making a scan through the various wavelengths.
2. Select a significant absorption peak which corresponds only to the compound of interest.
3. Next, prepare a series of standard solutions of the compound in question to cover the range of concentrations that is expected.
4. Determine the absorption values for each of these solutions to obtain the calibration plot (or the graph of the Beer–Lambert Law), which basically is a plot of absorbance versus concentration.
5. Place the sample of unknown concentration in the spectrophotometer, and measure the absorbance for this unknown solution.
6. Determine the concentration of the compound in question from the calibration curve.

If the molar extinction coefficient, ε, is known, the concentration can be calculated directly from the absorbance using the equation from the Beer–Lambert Law, without the calibration graph.

10.1 Determine the Concentration of the $[Ni(H_2O)_6]^{2+}$ Complex

The Task:

The complex, $[Ni(H_2O)_6]^{2+}$, absorbs electromagnetic radiation of a wavelength of about 580 nm. The amount of radiation that is absorbed by the complex is dependent on the amount of the complex present. You are supposed to plan an experiment to determine the concentration of $[Ni(H_2O)_6]^{2+}$ in an unknown solution with the following chemicals and apparatus:

- 25 g of solid $NiCl_2$;
- An unknown solution of $[Ni(H_2O)_6]^{2+}$ of about 0.5 mol dm^{-3};
- 5 cm^3 graduated pipette;
- 100 cm^3 volumetric flask;
- 20 cm^3 volumetric flasks;
- Spectrophotometer;
- Deionized water; and
- Standard glassware in the lab.

Q Where should we start thinking?

A: (i) What is the purpose of the plan?
— To determine the concentration of $[Ni(H_2O)_6]^{2+}$ in an unknown solution.

(ii) What do you need to know in order to determine the concentration of $[Ni(H_2O)_6]^{2+}$ in an unknown solution?
— We need to construct a calibration plot and from the graph, determine the concentration of $[Ni(H_2O)_6]^{2+}$ in an unknown solution.

(iii) How do you construct the calibration plot?
— We need to first make a standard solution of $[Ni(H_2O)_6]^{2+}$. From here, dilute the standard solution to get different solutions of different concentrations. Then, measure the absorbance of each of the diluted solutions and the unknown $[Ni(H_2O)_6]^{2+}$ solution. Draw the calibration plot and from the plot, use the absorbance of the unknown $[Ni(H_2O)_6]^{2+}$ solution to find out its concentration.

(iv) So, what is the concentration of the standard $[Ni(H_2O)_6]^{2+}$ solution that you should make?
— Since the unknown $[Ni(H_2O)_6]^{2+}$ solution has a concentration of about 0.5 mol dm^{-3}, it is reasonable to make a standard solution of concentration 1.00 mol dm^{-3} and from here carry out the dilution.

(v) How are you going to carry out the dilution?
— We could carry out a direct dilution by making concentrations of 0.8, 0.6, 0.4, and 0.2 mol dm^{-3}.

> **Q** Why can't we carry out a serial dilution instead?

A: Serial dilution is a stepwise dilution, and usually the dilution factor at each step is a constant, resulting in a geometrical progression of the concentration in a logarithmic fashion. For example, a two-time dilution could be 1.0 mol dm^{-3}, 0.5 mol dm^{-3}, 0.25 mol dm^{-3}, 0.125 mol dm^{-3}, etc. As the dilution required in this case does not have a constant dilution factor, we do not employ the serial dilution technique.

> **Q** Is serial dilution better than direct dilution?

A: Not necessarily. If the standard solution is not handled well or the mixing during dilution is not good, these errors would be compounded in subsequent serial dilutions, mixing, or transferring of solution.

Pre-Experimental Calculations:

Making a standard solution:

Assuming that the standard solution of $[Ni(H_2O)_6]^{2+}$ has a concentration of 1.00 mol dm^{-3}, to make 100 cm^3 of the standard solution:
Amount of $[Ni(H_2O)_6]^{2+}$ in 100 cm^3 of concentration of 1.00 mol dm^{-3}

$$= \frac{100}{1000} \times 1.00 = 0.100 \text{ mol}$$

Molar mass of $NiCl_2$ = 56.7 + 2(35.5) = 127.7 g mol^{-1}
Mass of $NiCl_2$ needed = 0.100 × 127.7 = 12.77 g.

Diluting the standard solution using direct dilution:

During dilution, the amount of particles before and after dilution is the same!

$$\text{Using } C_{diluted} \times V_{diluted} = C_{standard} \times V_{standard},$$

where $C_{diluted}$ = concentration of diluted solution,
$V_{diluted}$ = volume of diluted solution,
$C_{standard}$ = concentration of standard solution, and
$V_{standard}$ = volume of standard solution.

The volume of each diluted solution to be made is 20 cm^3.

For concentration of 0.8 mol dm^{-3}, $V_{standard}$ needed = $\frac{0.8 \times 20}{1.0}$ = 1.6 cm^3.

For concentration of 0.6 mol dm^{-3}, $V_{standard}$ needed = $\frac{0.6 \times 20}{1.0}$ = 1.2 cm^3.

For concentration of 0.4 mol dm^{-3}, $V_{standard}$ needed = $\frac{0.4 \times 20}{1.0}$ = 0.8 cm^3.

For concentration of 0.2 mol dm^{-3}, $V_{standard}$ needed = $\frac{0.2 \times 20}{1.0}$ = 0.4 cm^3.

The Procedure:

Procedure for making a standard solution:

(i) *Weigh* accurately *12.77 g* of the sample.
(ii) *Dissolve* the sample in about *30 cm^3* of water in a *beaker*.

(iii) Transfer the solution into the *100 cm³ volumetric flask* after ensuring that *all the solid has dissolved*. Rinse the beaker with water *a few times* and transfer *the washings* into the volumetric flask, to ensure *quantitative transfer*.
(iv) *Top up* the solution in the volumetric flask to the *graduation mark* using a dropper. Shake the flask well to get a *homogeneous solution*.

Procedure for direct dilution:

(i) Use a 5 cm³ *graduated pipette* to transfer *1.6 cm³* of the 1.00 mol dm⁻³ standard solution into a *20 cm³ volumetric flask*. *Top up* to the graduation mark using deionized water with a *dropper*. Shake the flask well to get a *homogeneous solution*.
(ii) *Repeat the dilution process* by using 1.2 cm³, 0.8 cm³, and 0.4 cm³ of the standard solution.

Procedure to obtain the calibration plot:

(i) *Measure the absorbance* of each of the solutions of concentration, 1.0, 0.8, 0.6, 0.4, and 0.2 mol dm⁻³, using a *spectrophotometer* set at a wavelength of *580 nm*.
(ii) *Plot a graph* of absorbance against the concentration values of the different $[Ni(H_2O)_6]^{2+}$ solutions.

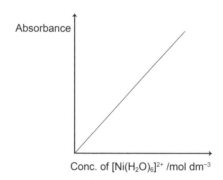

(iii) *Measure the absorbance* of the unknown $[Ni(H_2O)_6]^{2+}$ solution.
(iv) From the calibration plot, *determine the concentration value* of the unknown $[Ni(H_2O)_6]^{2+}$ solution.

10.2 Determine the Formula of the $[Ni(NH_3)_n]^{2+}$ Complex

The Task:

The complex, $[Ni(NH_3)_n]^{2+}$, absorbs electromagnetic radiation of a wavelength of about 650 nm. The amount of radiation that is absorbed by the complex is dependent on the amount of the complex present. You are supposed to plan an experiment to determine the value of n in the $[Ni(NH_3)_n]^{2+}$ complex with the following chemicals and apparatus:

- 1.00 mol dm^{-3} Ni^{2+}(aq) solution;
- 2.00 mol dm^{-3} NH$_3$(aq) solution;
- 10 cm^3 graduated pipettes;
- Spectrophotometer;
- Deionized water; and
- Standard glassware in the lab.

Q Where should we start thinking?

A: (i) What is the purpose of the plan?
— To determine the value of n in the $[Ni(NH_3)_n]^{2+}$ complex.

(ii) What do you need to know in order to determine the value of n in the $[Ni(NH_3)n]^{2+}$ complex?
— If we know the ratio of Ni^{2+} to NH$_3$, we would know the value of n.

(iii) How do you determine the ratio of Ni^{2+} to NH$_3$?
— Since the amount of radiation absorbed by the complex is dependent on the amount of the complex present, we can vary the ratio in which the Ni^{2+}(aq) and NH$_3$(aq) solutions are mixed and then measure the amount of absorbance of the solution. Thus, when the Ni^{2+}:NH$_3$ ratio is high, there would not be a sufficient amount of NH$_3$ to form the $[Ni(NH_3)_n]^{2+}$ complex; hence, the absorbance would be low. Similarly, with a low Ni^{2+}:NH$_3$ ratio, there would not be sufficient Ni^{2+} to form the complex; the absorbance value would be low as well. Therefore, the peak of absorption of the radiation

corresponds to the $Ni^{2+}:NH_3$ ratio which produces the highest amount of the $[Ni(NH_3)_n]^{2+}$ complex.

The Procedure:

(i) Use a different *graduated pipette* for each of the given solutions, to prepare nine sets of the following solutions in accordance to their volumes required:

Vol. of Ni^{2+}(aq)/cm³	0.0	1.0	2.0	3.0	4.0	5.0	6.0	7.0	8.0	9.0	10.0
Vol. of NH_3(aq)/cm³	10.0	9.0	8.0	7.0	6.0	5.0	4.0	3.0	2.0	1.0	0.0
Total vol./cm³	10.0	10.0	10.0	10.0	10.0	10.0	10.0	10.0	10.0	10.0	10.0
Measured Absorbance, A											

(ii) *Measure the absorbance* of each of the solutions using a *spectrophotometer* set at a wavelength of *650 nm*.

(iii) *Plot a graph* of absorbance against the volumes of Ni^{2+}(aq) used for each of the solutions.

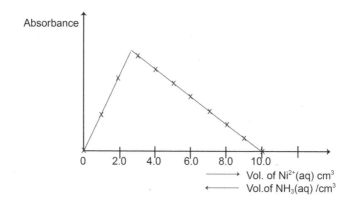

(iv) *Find the maximum point of absorbance* by drawing two straight lines which will intersect at a particular $Ni^{2+}:NH_3$ volume ratio.

Treatment of Results:

From the graph, the intersection point occurs at 2.5 cm^3 of 1.00 mol dm^{-3} Ni^{2+}(aq) solution and 7.5 cm^3 of 2.00 mol dm^{-3} NH$_3$(aq) solution:

Amount of Ni^{2+} present = $\frac{2.5}{1000} \times 1.00 = 2.5 \times 10^{-3}$ mol

Amount of NH$_3$ present = $\frac{7.5}{1000} \times 2.00 = 15.0 \times 10^{-3}$ mol

Therefore, mole ratio of Ni^{2+}:NH$_3$ = 2.5 × 10^{-3} : 15.0 × 10^{-3} = 1:6.

Hence, the value of n in [Ni(NH$_3$)$_n$]$^{2+}$ is 6.

Note: You can use the above planning process to determine the chemical formula of the metal ion to ligand ratio for any colored complex!

INDEX

absorbance, 191, 278
acid dissociation constant, 31
acid halide, 249
acid–metal, 222
acid spray, 17
activation energy, 184
adsorption, 268
alcohol, 248, 250
aldehyde, 249
aliquot, 171
alkene, 247
amide, 250
amine, 250
amphoteric, 216
analyte, 1, 55
anode, 198, 203
anodization, 214
appreciable hydrolysis, 234
Avogadro's constant, 199, 210
Avogadro's number, 205

barometer, 88
basic hydrolysis, 38
Beer–Lambert Law, 278, 279
Buchner filtration, 264
burette, 2

calibration plot, 278, 282
calorimeter, 142, 143, 145
capillary action, 265, 266
carboxylic acid, 249
cathode, 198, 203
cell potential, 200
charge density, 57
chemical energy, 197
chemical kinetics, 157
chromatogram, 266
chromatography, 265
colorimeter, 191
complex formation, 222
concentration gradient, 208
conjugate acid, 33
conjugate base, 33, 38
continuous method, 157, 158
copper purification, 212
crucible, 58

2,4-DNPH, 249, 251, 253
dessicator, 59
Devarda's alloy, 225
dibasic acid, 34
dilution, 6, 11
dilution factor, 280

direct dilution, 280, 282
displacement, 222
disproportionation, 222
distribution coefficient, 263
double-indicators method, 10, 14
downward displacement of water, 78
drying agents, 82

effervescence, 224
electrical energy, 197
electric cell, 197, 198
electrodes, 198
electrolysis, 197
electrolyte, 198
electrolytic cell, 197
electromagnetic radiation, 277
endothermic, 120
energy change, 103
energy cycle, 104
enthalpy, 104
enthalpy change, 112, 117
enthalpy change of combustion, 140, 148
enthalpy change of hydrogenation, 151
enthalpy change of neutralization, 137
enthalpy change of reaction, 132
ester, 250
exothermic, 113, 119

Faraday's constant, 199, 210
Faraday's First Law of Electrolysis, 199
Faraday's Second Law of Electrolysis, 199
Fehling's reagent, 249, 251
first-order reaction, 160
fractional distillation, 259, 260

fractionating column, 260, 261
frequency, 277
frictional heat, 107
frictionless syringe, 78
functional group, 247

gas collection method, 77, 164
Gibbs Free Energy, 198
graduated dropping funnel, 78
gravimetric analysis, 55
gravimetry, 74

half-equivalence point, 31, 33
half-life, 173
halogenoalkane, 248
heat capacity, 103, 144, 145
heat change, 103
heating mantle, 258
Henderson–Hasselbach equation, 33
Hess's Law, 112
hydrolysis, 247, 252

ideal gas constant, 90
independent variable, 159
initial rate method, 157, 158
instantaneous dipole–induced dipole, 48
intermolecular forces, 90
intramolecular covalent bonds, 57
iodine–starch complex, 47
ion-dipole interactions, 152, 262

ketone, 249

Law of Conservation of Energy, 112
Liebig condenser, 258, 260

mass concentration, 21
methyl orange, 1, 8, 10, 38
mobile phase, 268, 270

molar concentration, 21
molar extinction coefficient, 278, 279
molar gas constant, 187
molar volumes, 79
monochromatic, 277

nitrile, 250
nucleophile, 255
nucleophilic substitution, 255

organic synthesis, 257
oxidation, 197, 198

partial double bond, 255
partition coefficient, 50, 263
phenol, 248
phenolphthalein, 1, 8, 10, 26, 38
pipette, 2
position of equilibrium, 30
precipitate, 221

qualitative analysis, 221, 247
quantitative transfer, 7

rate constant, 173
rate equation, 159
rate method, 175
rate of reaction, 157
recrystallization, 271
redox titration, 38, 42
reduction, 197, 198
reflux, 258
retardation factor, 267
retention factor, 267

salt-bridge, 202
saturated solution, 274
second-order reaction, 160
selective discharge, 198

separating funnel, 263, 264
serial dilution, 280
simple distillation, 259
solubility, 68
solubility product, 26, 72
spectrophotometer, 278
spontaneous, 198
standard conditions, 203
standard enthalpy change of
    neutralization, 113
standard hydrogen electrodes, 200
standard reduction potential, 204
standard solution, 2, 6
stationary phase, 268, 269, 270

temperature gradient, 103
tertiary alcohol, 250
thermal dilution, 111
thermochemical experiment, 104, 110
thermometric titration, 126, 129,
    136
thermometry, 126
titer value, 5
titrand, 1
titrant, 5
titration, 1
titrimetric method, 158
Tollens's reagent, 249, 251

van der Waals' forces, 48
vapor pressure, 152
voltaic cell, 197, 200
volumetric flask, 6

wavelength, 277
weak acid–strong base titration, 26, 38
working range, 8, 38

zero-order reaction, 159